Stability and Control of Aircraft Systems

Stability and Control of Aircraft Systems
Introduction to Classical Feedback Control

Roy Langton

John Wiley & Sons, Ltd

Other Wiley Editorial Offices

John Wiley & Sons Inc., 111 River Street, Hoboken, NJ 07030, USA

Jossey-Bass, 989 Market Street, San Francisco, CA 94103-1741, USA

Wiley-VCH Verlag GmbH, Boschstr. 12, D-69469 Weinheim, Germany

John Wiley & Sons Australia Ltd, 42 McDougall Street, Milton, Queensland 4064, Australia

John Wiley & Sons (Asia) Pte Ltd, 2 Clementi Loop #02-01, Jin Xing Distripark, Singapore 129809

John Wiley & Sons Canada Ltd, 22 Worcester Road, Etobicoke, Ontario, Canada M9W 1L1

Wiley also publishes its books in a variety of electronic formats. Some content that appears in print may
not be available in electronic books.

Library of Congress Cataloging in Publication Data

Langton, Roy.
   Stability and control of aircraft systems : introduction to classical feedback control / Roy Langton.
     p.   cm.
   ISBN 0-470-01891-7
   1. Stability of airplanes.   2. Airplanes—Control.   I. Title
   TL574.S7L35 2006
   629.132′36—dc22

                                                 2006015974

British Library Cataloguing in Publication Data

A catalogue record for this book is available from the British Library

ISBN-13 978-0-470-01891-0 (HB)
ISBN-10 0-470-01891-7 (HB)

Typeset in 10.5/12.5pt Palatino by Integra Software Services Pvt. Ltd, Pondicherry, India
Printed and bound in Great Britain by TJ International, Padstow, Cornwall
This book is printed on acid-free paper responsibly manufactured from sustainable forestry in which at
least two trees are planted for each one used for paper production.

Contents

Series Preface

The field of aerospace is wide ranging and covers a variety of products, disciplines and domains, not merely in engineering but in many related supporting activities. These combine to enable the aerospace industry to produce exciting and technologically challenging products. A wealth of knowledge is contained by practitioners and professionals in the aerospace fields that is of benefit to other practitioners in the industry, and to those entering the industry from University.

The Aerospace Series aims to be a practical and topical series of books aimed at engineering professional, operators, users and allied professions such as commercial and legal executives in the aerospace industry. The range of topics spans design and development, manufacture, operation and support of aircraft as well as infrastructure operations, and developments in research and technology. The intention is to provide a source of relevant information that will be of interest and benefit to all those people working in aerospace.

Ian Moir, Allan Seabridge and Roy Langton

Preface

There are many textbooks on the subject of feedback control; however, most are highly mathematical and, as a result, often repel the seasoned engineer who may have become a little rusty regarding the rigors of certain aspects of mathematics that include such things as differential equations and complex number theory. In a similar manner, unnecessarily complex mathematics can be a turnoff to engineering students who might otherwise find the control systems engineering field both challenging and exciting.

This book is not a textbook in the traditional sense but an attempt by the author to give back to the next generation of control systems engineers a guidebook containing easy to follow descriptions of the important aspects of classical control supported by examples based on real world events that have occurred during the author's career in the aerospace industry. The arrangement and content of the book is an attempt to provide an effective answer to the question 'What would have been most useful to me as a prospective systems engineer in the pre-to-post graduate timeframe seeking guidance and insight into the fundamentals of feedback control?'.

In the opinion of the author, complex mathematics need not be a significant barrier to learning if a pragmatic presentation methodology can be developed providing a more straightforward approach to the subject that can be more easily absorbed by the practicing engineer and that provides an inspiration to the prospective control engineering graduate.

In the current world of increasing complexity and functional integration in all areas of engineering and technology, the engineer who did not take the course on 'stability and control' operates at a serious disadvantage. This is a common issue amongst many older mechanical engineering graduates because historically control theory has been a part of the electrical engineering curriculum. Even though this is becoming less typical in most learning institutions the seasoned engineer with a mechanical engineering background has, more often than not, never been exposed to the subject of feedback control theory. An additional problem with academia is the fact that much of the material taught is not in common practice within the industry. As a result the learning experience becomes more of a mathematical exercise that misses out many of the pragmatic methods that have been established as most effective in the design and development departments of industry.

Today's engineers are required more and more to be both specialists in their area of expertise and generalists who understand the complete functional context of the application where their products are being used. Also, many of today's products contain multiple engineering technologies. What once were single discipline mechanical, hydraulic or pneumatic products and systems now contain integrated electronic sensors and, in many instances today, software. Control theory reduces these widely varied technical disciplines into their important dynamic characteristics expressed as transfer functions from which the subtleties of dynamic behavior can be analyzed and understood.

The objective of this introductory book on feedback control is developed around the generic closed loop control system concept illustrated by the diagram of Figure 0.1: As shown, the typical system comprises a

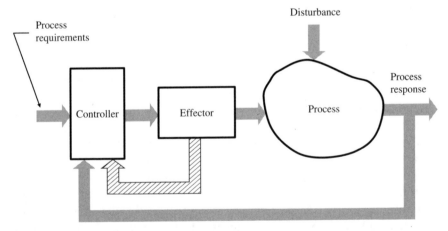

Figure 0.1 Generic feedback control system

control element, an effector and a process to be controlled. The process requirements are compared with the process response in the controller whose task is to generate actions that ultimately bring the process in line with the required state. The effector represents a power amplification stage or 'muscle' that takes the control output signals and converts them into a form that can be used to effect a change in the process. The process may be any number of things from a simple actuator to a major aircraft control system. The arrow connecting the process response to the controller represents the feedback of process states to the controller, hence the term feedback loop. The effector may also incorporate feedback to the controller as indicated by the shaded arrow.

An important feature of the feedback control system is that external disturbances which affect the response of the process will be sensed, and ultimately compensated for, by the ensuing corrective action determined by the controller. The challenge for the control system designer is to establish the best control algorithm that will provide the optimum performance in terms of accuracy, dynamic response and stability. The objective of this book is to provide the reader with the basic tools to understand the design processes and to visualize the functional behavior associated with feedback control systems.

As part of the aerospace series the material presented in this book is related to aircraft control system situations almost exclusively. Furthermore, the extensive background of the author in the areas of flight controls, hydraulics, fuel and engine control systems forms the basis for many of the design examples and reinforcement exercises developed.

At this point it is appropriate for the reader to recognize that aircraft closed loop control systems vary substantially in their criticality and response needs. While the primary focus of this book on stability and control is aimed at the tightly coupled fast response systems where stability and response requirements are important design and operational issues often with demanding specifications, there are many interactive control systems within a typical aircraft that are much more loosely coupled but nevertheless must be recognized and evaluated from a response and performance perspective. Figure 0.2 illustrates this point by showing schematically the functional relationship between the various layers of functionality associated with the control of a modern aircraft.

Shown in Figure 0.2 are a number of nested control loops with control surface actuators at the center which determine the immediate aircraft response and attitude. At the second level is the flight director/autopilot

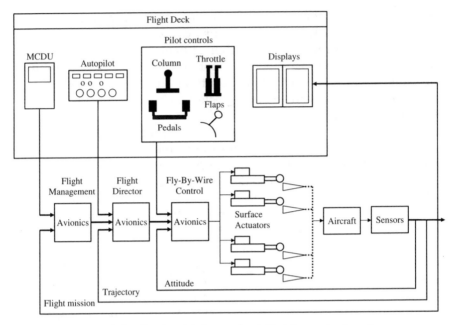

Figure 0.2 Typical integrated flight control system

control loop which determines the trajectory of the aircraft within its local airspace. Around these two control loops is the navigation system which controls the aircraft's mission through space relative to Earth coordinates.

The outer loops become less tightly coupled but they are by definition closed loop control systems and as such must be recognized by the control systems engineer in terms of response needs and potential interaction between the various control layers. These comments are presented here to provide a perspective and awareness to the reader of the complexity of the modern aircraft in terms of stability and control and to be cognizant of the additional 'outer' control loops that are invariably involved in the overall performance of the aircraft.

While the intent of this book is to minimize the mathematical content, all of the key analytical procedures are developed from first principles in the interest of completeness and to satisfy the reader with a strong interest in the mathematics. The underlying methodologies, graphical aids and guidelines described, however, can be developed using fairly simple algebraic principles that are intended to provide the practitioner with a good 'feel for the problem'.

Part of the fascination of understanding the principles of feedback control is learning to be able to appreciate the functional behavior of control systems through the interpretation of simple block diagrams and to appreciate the fact that the dynamic functionality of complex integrated systems is typically not intuitive.

The author has a long-standing belief that a basic understanding of feedback control systems design, analysis and testing allows the design/development engineer to have a clearer understanding of the dynamic functional behavior that results when multiple components are combined within a system to form an integrated functional entity.

The end objective of this book, therefore, is in effect to 'switch on the light' in the dark room of system design and development for those engineers who missed the opportunity to complete formal training in control theory at university. This will provide the know-how necessary to minimize problems with fielded systems in the area of operational performance thus increasing the readers' effectiveness in the eyes of both their employers and their customers.

Roy Langton

1

Developing the Foundation

Classical feedback control theory is an inherently mathematical subject and as such can scare off the would-be practitioner if presented in rigorous mathematical form. While there is no way to fully absorb and apply the basic concepts of feedback control without any mathematics it is definitely possible to ease into the subject matter gradually thus allowing the reader to pursue with some curiosity, and hopefully some excitement, the subtleties of closed loop system behavior.

This chapter attempts to provide the reader with a basic understanding of the terminology associated with control theory and through the use of simple examples to familiarize the reader with some of the key mathematical tools needed to apply feedback control principles to engineering problems. To begin with we need to discuss the issue of engineering units since there are fundamental differences between the standards in use at teaching institutions and within industry both in Europe and the United States. This is followed by a discussion regarding the use of block diagrams as a way to describe the functionality of closed loop control systems. Differential equations describe the dynamic behavior of physical systems and are at the core of feedback control theory. Here, however, they are transformed from the mathematics domain into an easy to assimilate block diagram form using simple examples.

Stability and Control of Aircraft Systems: Introduction to Classical Feedback Control   R. Langton
© 2006 John Wiley & Sons, Ltd

Finally, a brief refresher on the subject of complex number theory is presented since there is no way to avoid this aspect of mathematics in describing oscillatory behavior which is such an important part of the process for defining closed loop system response and stability to be developed and discussed later. Throughout this chapter, and also throughout the rest of this book, every attempt is made to use simple examples to reinforce the learning process and to present the material in an easy-to-follow manner using a minimum of complicated mathematics.

1.1 Engineering Units

An important aspect of understanding control systems involves verifying the meaning of equations by analyzing the units associated with the various constants and variables used. For completeness two unit standards are addressed here in order to represent adequately the academic and industrial backgrounds of readers from both Europe and the United States.

In Europe the International System of Units (SI) has been adopted as the standard in both academia and industry, while in the United States, the SI system taught almost exclusively in universities is by no means the standard of industry where the US/Imperial standards are still used by the majority. Perhaps the most significant difference between the SI and US/Imperial systems is regarding the treatment of mass and inertia and their relationships with linear and angular acceleration. The following section addresses this aspect of the two units systems before comparing and reconciling them into a common method of understanding.

1.1.1 International System of Units (SI)

In this system the kilogram (kg) is used to define the mass of an object representing its inertial resistance to acceleration. Force is expressed in 'newtons' (N) and acceleration in meters per second^2 (ms^{-2}).

Considering the basic equation:

$$\text{force} = \text{mass} \times \text{acceleration} \ (F = m \times a).$$

The SI system requires that a force of 1 newton applied to a mass of 1 kilogram will result in an acceleration of 1 meter per second^2.

For the equation above to balance, the units of newtons must be equivalent to $\mathrm{kg\,m\,s^{-2}}$, that is:

$$\mathrm{kg\,m\,s^{-2} = kg \times m\,s^{-2}}.$$

Using the same logic we can express the units of kilogram mass as:

$$\mathrm{kg = N/m\,s^{-2}}.$$

Thus we have expressed mass as a force per unit acceleration.

1.1.2 US/Imperial Units System

The imperial units system (adopted with a number of modifications by the United States) developed from the fact that the local gravitational force is always the same and equal to:

$$g = 32.2\,\mathrm{ft/sec^2} \qquad (\mathrm{or}\ 386.4\,\mathrm{in./sec^2}).$$

The basic equation: force $=$ mass $\times$ acceleration is reconciled by defining mass as weight divided by g so that:

$$F = (W/g) \times \mathrm{acceleration}.$$

Here the unit-balance equation is:

$$\mathrm{lb = lb/ft\,sec^{-2} \times ft\,sec^{-2}}.$$

For most aerospace engineering problems inches (in.) are used for displacement rather than feet (ft) for obvious reasons of scale.

An alternative method widely in use is to express the mass term as pounds mass ($\mathrm{lb_m}$) and force as pounds force ($\mathrm{lb_f}$); however, for simplicity we will adopt the convention where pounds (lb) always represents force and mass is expressed as the inertial force resisting acceleration in lb per unit acceleration, i.e. $\mathrm{lb/(ft\,sec^{-2})}$.

1.1.3 Comparing the SI and US/Imperial Units Systems

Going back to the basic equation the SI system denotes:

$$\text{force (N)} = \text{mass (kg)} \times \text{acceleration (m s}^2).$$

We can also express this equation as follows:

$$\text{force (N)} = \text{inertial resistance (N/m s}^2) \times \text{acceleration (m/s}^2).$$

This is now in the same form as the US/Imperial representation, that is:

$$\text{force (lb)} = \text{inertial resistance (lb/ft sec}^{-2}) \times \text{acceleration (ft/sec}^2).$$

This same approach can be readily applied to the rotational equivalent of the above linear examples where:

$$\text{torque} = \text{polar (rotational) inertia} \times \text{rotational acceleration.}$$

The units equations for the SI and US/Imperial systems are as follows, respectively:

$$\text{N m} = \text{N m/(r/s}^2) \times \text{r/s}^2 \text{ and lb ft} = \text{lb ft/(rads/sec}^2) \times \text{rads/sec}^2.$$

The intent of the above is to point out that both unit standards can be expressed in a similar manner. The interpretation above is used throughout this book because it is considered to be the most intuitive to the reader.

1.2 Block Diagrams

Block diagrams are used extensively by control systems engineers to provide a visual insight into the functionality of closed loop control systems; it is therefore important for the reader to get comfortable with this concept. Block diagrams are a convenient way to depict the various elements of a control system as a number of interconnected boxes or blocks wherein the input to a box/block multiplied by the contents of the box/block defines the output from that box/block.

The contents of blocks may contain dynamic terms or they may be scalar quantities such as gains, ratios, performance coefficients, etc.. In the process of calculating the output from a block, the units of the input term and the block contents are also multiplied together to define the units of the output from that block. For example, Figure 1.1 shows the relationship between a fluid valve displacement (the input to the block X in millimeters (mm) and the output volume flow from the valve Q in liters per second (l/s). The term inside the block, the valve gain K_V, has units of liters per second per millimeter of valve displacement (l/s)/mm.

Another important feature used in block diagrams is the summation device which allows two or more signals to be added or compared. The diagrammatic representation of a summation device is shown in Figure 1.2. Here the output from the summation device can be either the sum or the difference of the inputs to the device depending upon the signs written alongside the arrows of each input. Also, as indicated by the Figure 1.2, a summing device can have multiple inputs. An important point to remember when using summing or comparison devices is that the inputs and outputs to/from the device must all have the same units. Using a summing device to define the difference between two variables is a fundamental aspect of feedback control where the required output from a control system is compared with the actual (measured) output and the difference used to cause the output to move towards the desired value.

1.2.1 Examples of Summation (or Comparison) Devices

The most obvious example of a comparison device taken from our everyday experience is the thermostat used to control the tempera-ture in our home or office. This system is called a 'bang–bang' control

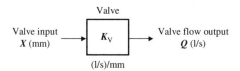

Figure 1.1 Fluid valve block diagram example

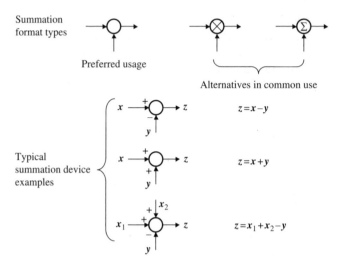

Figure 1.2 Summation device representation

since it is either 'on' or 'off'. If the room temperature falls below the desired value, the thermostat switches the heating system 'on' and when the temperature reaches the desired value the heating system is switched 'off'. (There is usually some hysteresis between the 'on' and 'off' settings to prevent 'chatter' about the set point.) This type of system is 'nonlinear' because of the on/off or discontinuous characteristic of the controller. For the most part we shall be dealing with 'linear' systems where the action taken by the control system is in proportion to the magnitude of the difference between the desired value and the actual output. There are many examples of summation/comparison devices used in all engineering disciplines including mechanics, hydraulics, pneumatics and electronics. Presented below are just a few of the more common types.

The Float Control Valve

Everyone is familiar with the float mechanism inside the toilet cistern that limits the level of water in the cistern during refill following the flush action. Figure 1.3 shows this system both schematically and in block diagram form. This same concept is used extensively in aircraft fuel systems where float valves are used to provide a signal to the refueling or fuel transfer system that the fuel tank is full. In this case the

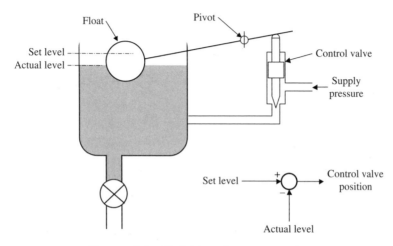

Figure 1.3 Fluid level comparator

float valves do not act directly on the flow control valve but provide a
servo pressure signal that initiates control valve action shutting off flow
to that tank. For this reason these valves are termed 'float pilot valves'.

Mechanical Linkage Summing

Mechanical linkage summing is an extremely simple form of summing
device that is used extensively in many aircraft flight control systems
in service today. Movements of the pilot's control column are typically translated via cables and pulleys or push–pull rods to mechanical inputs to servo control actuators at the control surface. These
actuators provide the necessary muscle to overcome the aerodynamic forces associated with high speed flight. The pilot's input is
compared with the control surface position (servo actuator output) in
a mechanical summing linkage arrangement similar to that shown in
Figure 1.4. In a similar manner inputs from an autopilot actuator can
be summed with the pilot's command to provide an auto-stabilization
function.

The Speed Governor

The speed governor goes back more than 200 years to James Watt who
invented the 'flyball governor' as a mechanism to control the speed of
a steam engine that did not require human intervention. Derivatives of

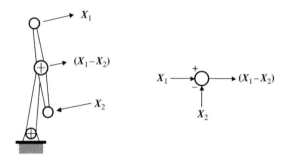

Figure 1.4 Mechanical linkage summer/comparator

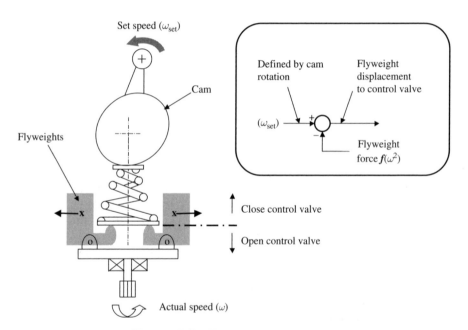

Figure 1.5 Flyweight governor

this device are used throughout industries using rotating machinery. Figure 1.5 shows such a device in schematic and block diagram form. Today, flyweights rather than balls are more typical offering a more compact design. The basic concept, however, remains unchanged. The rotation from the machinery to be controlled causes the weights to be thrown outward due to the centrifugal force. This force is compared with an opposing spring force defining the desired operating speed. If the speed exceeds this value the resulting upward motion of the flyweights is used to close the valve supplying the energy to the rotating machinery under control. The square law relating speed and flyweight force is non-linear but a linear approximation can be used effectively in control systems analysis. In some sophisticated devices associated with gas turbine fuel controls, the flyweight governor is integrated with a non-linear servo mechanism that is designed to compensate for the square law effect resulting in an output displacement directly proportional to the rotational speed. With the advent of digital electronic engine controls this non-linear problem is easily taken care of using software.

The Operational Amplifier

The 'op-amp' is the cornerstone of analog electronic circuitry and can be used to sum or difference two or more voltages. Figure 1.6 shows a schematic of an operational amplifier with three separate inputs connected to the summing junction of the amplifier via identical resistors R. The equivalent block diagram is also shown in this figure. The output of the amplifier is also shown connected to the same junction via a resistor of the same value. Since the voltage gain of the amplifier is very high (say $>10^5$) we can treat the summing junction for all three inputs and the feedback as a 'virtual ground' and so the current

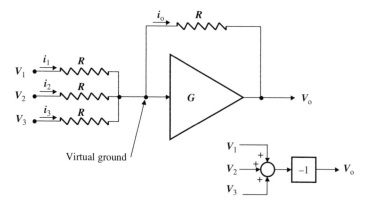

Figure 1.6 The summing amplifier

flowing into the amplifier can be assumed to be zero. Therefore by Kirchoff's Law the sum of the currents from the three inputs into the virtual ground must equal the current flowing out through the feedback resistor, i.e.

$$\frac{V_1}{R} + \frac{V_2}{R} + \frac{V_3}{R} = -\left(\frac{V_0}{R}\right).$$

The negative output is inherent in the hardware since the current flow in the feedback resistor is away from the summing junction and hence V_0 must be negative. Thus we have a simple summing amplifier. It is also possible to select different impedances for each input as well as the feedback to generate complex dynamic transfer functions.

Force Summing Bellows

Another commonly used summing device uses bellows and a summing link to generate a displacement proportional to two (or more) pressures in both hydraulic and pneumatic systems. Figure 1.7 shows such a device used to modulate the opening of a flapper valve as part of a servo mechanism. This arrangement is ideal since the servo flapper displacement is extremely small and therefore any error induced by the spring rate of the bellows will be negligible thus providing accurate force summing.

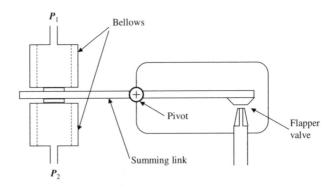

Figure 1.7 Force summing bellows arrangement

1.3 Differential Equations

This section reviews the differential equation which is the standard mathematical approach to defining the dynamic behavior of physical systems. Here we will limit our attention to linear differential equations with constant coefficients leaving how to deal with nonlinearities until later in the book. We will also describe how to express these equations in a simplified form that is readily adaptable to the block diagram concept previously discussed.

Let's begin with an equation that we are all familiar with. This equation is depicted schematically in Figure 1.8 and is, in fact, a second-order differential equation whose solution allows us to determine the position of the mass at any point in time. In more rigorous terms we can write the same equation in the following form:

$$F = M \left[\frac{d^2 x}{dt^2} \right] \text{ or } F = M\ddot{x},$$

where F is the applied force, M is the mass expressed as an inertial force per unit of acceleration, and $\frac{d^2 x}{dt^2}$ or $\ddot{x}$ is the second derivative of x, the position of the mass with respect to time.

Both the differentiation process, i.e. the calculation of the rate of change of a variable with respect to time,[1] and its inverse the integration process, are fundamental building blocks of feedback control theory and the 'D' notation (see below) is used to facilitate the expression of differential equations in block diagram form.

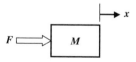

Figure 1.8 Schematic diagram of $F = M \times a$

---

[1] While differentiation (or integration) can be with respect to any variable, in feedback control it may be assumed that it is always with respect to time.

1.3.1 Using the 'D' Notation

Here 'D' means the derivative with respect to time, therefore:

$$\frac{dx}{dt} = Dx \text{ or } \dot{x}$$

and similarly:

$$\frac{d^2x}{dt^2} = D^2x \text{ or } \ddot{x}.$$

The integration process can now be expressed as $\frac{1}{D}$ since it is the inverse of the differentiation process.

The conventional expression for the integral of a function with respect to time is:

$$\int_{t_1}^{t_2} f(t)dt,$$

where the limits t_1 and t_2 define the span of time $(t_2 - t_1)$ over which the integration process takes place. When written without limits it can be inferred that the time span is from $t = 0$ to $t = \infty$. Therefore we can define the integral of the acceleration of x with respect to time as:

$$\int \left(\frac{d^2x}{dt^2}\right) dt = \frac{1}{D}\ddot{x} = \dot{x}$$

which is velocity, or, in block diagram form, as shown in Figure 1.9(a). Similarly the double integral of acceleration:

$$\iint \left(\frac{d^2x}{dt^2}\right) dt$$

can be expressed in block diagram form as shown in Figure 1.9(b) and this may be further simplified as indicated by Figure 1.9(c).

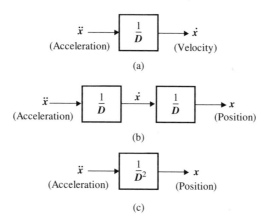

Figure 1.9 *D* notation examples using block diagrams

There is an important caveat that should be mentioned here as it relates to the notation described above. The output from the integration process requires a 'constant of integration' which, in the above examples is assumed to be zero. In other words one cannot compute the position x without knowing the initial velocity and position at the start of the integration process. When analyzing closed loop systems the default condition assumed above, is that the output of each integrator at time $t = 0$, is zero. This is usually acceptable but should be noted and understood.

The next step is to revisit the basic equation ($F = M \times a$) and to express it in block diagram form using the notation described above. The resulting block diagram is shown in Figure 1.10. As we learned from the block diagrams section above, the input to each box is multiplied by (or operated on by) the contents of the box to obtain the output from that box. When the term inside the box contains dynamic terms the output from the block is referred to as the 'response' and the contents of the block as the 'transfer function'.

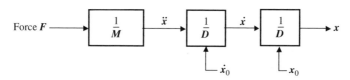

Figure 1.10 Block diagram of $F = M \times a$

Figure 1.10 gives a much clearer picture of what is happening than the general expression used at the outset. In order to calculate values of velocity and position one needs to know the initial conditions at the output of each integrator depicted above by the '0' subscripts for each integrator.

1.4 Spring–Mass System Example

In order to reinforce what we have learned so far we will apply these concepts to a more complex dynamic system. Consider now the spring–mass arrangement shown in Figure 1.11. From the original equation

$$\text{force} = \text{mass} \times \text{acceleration}$$

we can say:

$$(x_i - x_o)K - fDx_o = MD^2x_o.$$

At this point the control engineer would develop a block diagram representation of the system to provide a visual insight into its functionality. To accomplish this we rearrange the above force balance equation to put the highest derivative term (in this case D^2x_o) on the left-hand side as follows:

$$D^2x_o = \frac{1}{M}\left[(x_i - x_o)K - fDx_o\right].$$

It is now an easy task to construct a block diagram by summing all the terms in the equation to give D^2x_o and then integrating to give Dx_o

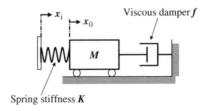

Figure 1.11 Spring–mass system

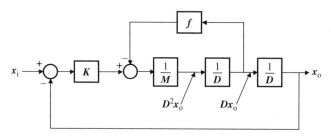

Figure 1.12   Spring–mass system block diagram

(velocity) and again to give x_o (position). Figure 1.12 shows the block diagram of this system developed this way.

This block diagram shows the spring–mass arrangement as a mini system with two feedback paths. The damping term fDx_o subtracts from the accelerating force and x_o is fed back and compared with x_i to determine the compression of the spring. This simple example demonstrates how block diagrams using the D notation to define the dynamic terms provide a more easy-to-understand way to express differential equations.

1.4.1 The Standard Form of Second-order System Transfer Function

In order to make the next point regarding second-order systems we need to revisit the original force balance equation. From this we will develop the standard representation of second-order systems that will prove to be extremely useful in the dynamic analysis methods to be developed later. Restating the spring-mass force balance equation and rearranging we obtain:

$$(x_i - x_o)K - fDx_o = MD^2x_o$$

or:

$$Kx_i = x_o(MD^2 + fD + K).$$

Dividing the above rearranged equation by the spring stiffness, K, we obtain the following equation relating the input and output displacements

$$\frac{x_o}{x_i} = \frac{1}{\left(\dfrac{M}{K}\right)D^2 + \left(\dfrac{f}{K}\right)D + 1}.$$

The right-hand side of this equation can be expressed in the following standard form:

$$\frac{1}{\left(\frac{D^2}{(\omega_n)^2}\right) + \left(\frac{2\zeta D}{\omega_n}\right) + 1},$$

where ω_n is the 'undamped natural frequency' in radians per second. In this specific case it is equal to $\sqrt{\dfrac{K}{M}}$ and ζ is the damping ratio defined as $\dfrac{f\omega_n}{2K}$ which has no units.

Let us check the units in each unit convention.

SI units:

$$K/M = \mathrm{N/m}/(\mathrm{N/m\,s^{-2}}) = \mathrm{s^{-2}} = \text{radians/second}^2 \qquad \checkmark$$

$$f\omega_n/2K = (\mathrm{N/m\,s^{-1}})\,(\mathrm{s^{-1}})/(\mathrm{N/m}) \dots \text{all units cancel} \qquad \checkmark$$

US/Imperial units:

$$K/M = (\mathrm{lb/in.})/(\mathrm{lb/in.\,sec^{-2}}) = \mathrm{sec^{-2}} = \text{radians/second}^2 \qquad \checkmark$$

$$f\omega_n/2K = (\mathrm{lb/in.\,sec^{-1}})\,(\mathrm{sec^{-1}})/(\mathrm{lb/in.}) \dots \text{all units cancel} \qquad \checkmark$$

The point of this standard expression for second-order systems is that it is independent of discipline, i.e. a second-order system associated with an electronic circuit, a chemical reaction, a hydraulic or pneumatic pressure oscillation can all be expressed in the standard form. This is an interesting aspect of feedback control systems in that it all boils down to a series of transfer functions that can be studied in control systems terminology and independent of any specific engineering discipline.

Here are some observations we can make regarding second-order system response. If the damping ratio goes to zero, the 'D' term in the transfer function also goes to zero implying that the system will oscillate at the undamped natural frequency forever, i.e. the oscillations will not decay. In the real world there is always some damping present, however small, causing the oscillations to die away to zero eventually. As the value of the damping ratio increases, the number of oscillations

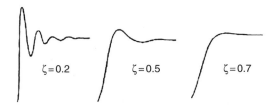

Figure 1.13 Spring–mass system response for different damping ratios

to damp out following a disturbance become fewer to the point where for $\zeta = 0.707$ there will be essentially no oscillations. To illustrate this point, Figure 1.13 shows the response of x_o to a sudden change in x_i for different values of damping ratio.

For a damping ratio of 1.0 the system is said to have 'critical damping' and the response equation for the system can be defined as the product of two identical first order terms as follows:

$$\frac{x_o}{x_i} = \frac{1}{\left(\frac{D}{\omega_n}+1\right)\left(\frac{D}{\omega_n}+1\right)}.$$

Multiplying out the two first-order factors yields:

$$\frac{D^2}{(\omega_n)^2} + \frac{2D}{\omega_n} + 1$$

implying from the above standard definition that $\zeta = 1.0$.   ✓

Let us look at the implications of the above in mathematical terms:

- For values of ζ between 0 and 1 the solution of the second-order differential equation will have imaginary terms.
- For the unique situation where $\zeta = 1$ the solution is represented by two equal real roots with a value of $-\omega_n$.
- As ζ increases above 1.0, the roots of the equation will be real but different with the difference becoming larger as ζ increases.

From the analysis of the second-order system it becomes clear that in the region of interest, i.e. when the response becomes oscillatory, the roots of the differential equation are imaginary numbers involving the square root of minus one designated by j. The next step therefore is to refresh our memories regarding the properties of complex numbers.

1.5 Primer on Complex Numbers

Complex numbers are necessary to allow us to express the square root of negative numbers which occur when solving differential equations such as we have shown above. This is particularly important in working with feedback control systems because it is when these systems exhibit oscillatory behavior or become unstable that the solutions involve the square root of negative numbers.

The concept is to define numbers as having both real and imaginary properties; real numbers being, therefore, just a special case where the imaginary part is zero. This can be expressed graphically by defining 'the complex plane' where the real part of a number is defined along the horizontal axis and the imaginary part, expressed as a product of the 'square root of (-1)' or j as indicated in Figure 1.14.

Referring to this figure we can define the complex number represented by point P in the complex plane so that:

$$P = a + jb.$$

Point P can also be considered as a vector P represented as the arrow between the origin and point P. This vector can be described as having a

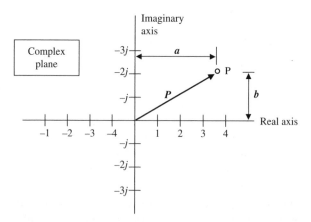

Figure 1.14 Graphical representation of the complex plane

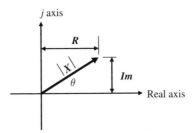

Figure 1.15 The polar coordinate form of a vector

magnitude equal to the length of the vector and a direction represented by the angle of the arrow relative to, say, the real axis. This method of describing a complex number is called the 'polar coordinate form' (see Figure 1.15).

Here we refer to the magnitude of the vector as the 'modulus' and the direction of the vector as the 'phase angle'. Referring to Figure 1.15, the modulus of the vector, $|x|$, is the length or magnitude of the vector and can be calculated via Pythagoras as:

$$|x| = \sqrt{R^2 + Im^2}.$$

The phase angle, θ is obtained from

$$\tan \theta = \frac{Im}{R}.$$

The above polar representation is fundamental to the methods used to analyze and understand the behavior of closed loop systems which we will develop in the forthcoming chapters.

1.5.1 The Complex Sinusoid

Sine waves are one of the most common ways to apply a disturbance to a system in order to understand the behavior of that system under dynamic conditions. Sinusoidal excitation is easy to apply in practice and by using a range of frequencies of excitation, the dynamic characteristics of a system can be evaluated over a specific frequency range of interest.

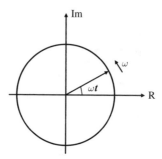

Figure 1.16 Rotating vector in the complex plane

From an analytical perspective, therefore, we need to familiarize ourselves with the way we can apply sinusoidal disturbances to the transfer functions of a system and to analyze the responses that result. Consider now the vector in Figure 1.16 rotating around the axis of the complex plane with an angular velocity of ω radians per second. If we were to plot the projected length of this vector onto the real axis against time we would obtain a sine wave. The interpretation of this arrangement is that a rotating vector in the complex plane appears in the 'real' world as a sinusoidal oscillation whose frequency is $\omega/2\pi$ cycles per second or 'hertz'. This is fairly intuitive and easy to understand.

Referring to the Figure 1.16 we can see that at any time 't' the phase angle of the vector with respect to the real axis is equal to ωt radians. Also, for a vector modulus of unity, the coordinates of the vector as it rotates around the origin are always:

$$\cos \omega t + j \sin \omega t.$$

Those readers who remember their high school mathematics will realize that this expression can be represented in polar coordinate form by the term $e^{j\omega t}$ which is referred to as '*the Complex Sinusoid*'. The application of this concept will be developed in the section on frequency response analysis where $e^{j\omega t}$ is used to describe the sinusoidal forcing function.

For the interested reader, the proof that the rectangular coordinate representation of a sine wave $\cos \omega t + j \sin \omega t$ is equivalent to the complex sinusoid (polar coordinate) representation $e^{j\omega t}$ is developed below: First we have to express the sine and cosine expressions as a series. The Taylor series for these are as follows:

$$\cos{(\omega t)} = 1 - \frac{(\omega t)^2}{2!} + \frac{(\omega t)^4}{4!} - \frac{(\omega t)^6}{6!} + \cdots$$

Prove this to yourself by plugging in a value for ωt in the above equation. For example, inserting $\omega t = \pi/3$ radians (60 degrees) will yield 0.5 which is the correct answer.

Similarly:

$$\sin{(\omega t)} = \frac{(\omega t)}{1!} - \frac{(\omega t)^3}{3!} + \frac{(\omega t)^5}{5!} - \cdots$$

If we multiply the sine series by j and add it to the cosine series we obtain the following series for $\cos \omega t + j \sin \omega t$:

$$1 + j(\omega t) - \frac{(\omega t)^2}{2!} - \frac{j(\omega t)^3}{3!} + \frac{(\omega t)^4}{4!} + \frac{j(\omega t)^5}{5!} + \cdots$$

This, believe it or not, is the Taylor series for $e^{j\omega t}$.

From the above we can also determine the following expressions:

$$\cos{(\omega t)} = \frac{1}{2}\left(e^{j\omega t} - e^{-j\omega t}\right)$$

and

$$\sin{(\omega t)} = \frac{1}{2j}\left(e^{j\omega t} - e^{-j\omega t}\right).$$

It is not essential to remember the above mathematical proofs to understand and analyze feedback control systems. It is presented here, however, for completeness.

1.6 Chapter Summary

This chapter has armed the reader with the basic tools that are needed to understand, analyze and evaluate the dynamic behavior of closed loop control systems. By introducing the concepts of the block diagram and the D operator, the dynamic behavior of physical systems can now be represented in a straightforward visual form where the contributions of the various elements of a system can be more clearly appreciated.

Perhaps the most challenging aspect covered so far is how oscillatory systems, such as the spring–mass example presented here, have roots that involve complex numbers and that the larger the imaginary part relative to the real part, the more oscillatory the system becomes. The standard form of second-order system where the transfer function is expressed as a function of the undamped natural frequency and damping ratio allows us to visualize better the meaning of these imaginary roots and how they relate to the physical behavior of the system.

An important point to take away from this chapter is that feedback control theory is essentially independent of discipline. All physical systems can be described dynamically by transfer functions. Once in this form the dynamic response attributes can be analyzed and evaluated as we shall see in the forthcoming chapters.

2

Closing the Loop

This chapter introduces the concept of the 'closed loop system' in general terms explaining the basic principles and issues involved. The significance of analyzing the open loop performance of a system to determine the degree of stability that can be achieved during closed loop system operation is explained. Various types of response testing of closed loop systems are discussed and the analytical techniques developed focus primarily on frequency response as the most popular method used by the control engineering community. Real world examples are used throughout to reinforce the concepts as they are developed.

2.1 The Generic Closed Loop System

A generic closed loop system (see Figure 2.1) can be described as a means of controlling the output of a process by comparing what is required with the actual output and using the output from this comparison to generate controlling actions that change the process output towards what is required. Note that in this simplified depiction, the effectors (muscle) can be assumed to be integrated within the controller block.

In the diagram, the input to the system (what is required) is compared with the feedback, which is a measure of the actual output, to establish the 'error'. Also shown in the diagram is a disturbance input representing external changes from outside the control loop that may affect the process. An example of a disturbance input in an electrical generator

Stability and Control of Aircraft Systems: Introduction to Classical Feedback Control   R. Langton
© 2006 John Wiley & Sons, Ltd

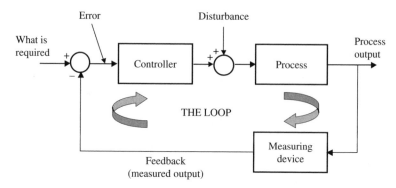

Figure 2.1 The generic closed loop system

speed control system would be a change in demand from the user of the electrical power. This would result in a sudden increase in generator torque causing a reduction in speed. The controller, sensing the speed reduction would then increase the drive torque to bring the speed back towards the set point. In aircraft or engine control systems, changes in air pressure and temperature as the aircraft changes speed and altitude may also be considered as disturbances that would require adjustment by the various control systems on board.

The controller must now use the error signal to create an input to the process such that the process output changes until the error is reduced (ideally to zero). Intuitively it can be seen that for the error to be very small, the controller must be 'sensitive' i.e. small errors must be capable of generating significant response if the control of the process is to be effective. In other words the controller must have a 'high gain'. Not so intuitive is the fact that with high gain closed loop systems, time delays around the loop can cause the system to be 'unstable'. Understanding the concept of 'stability' and having the tools to design closed loop systems that are well behaved from a stability and performance perspective is what 'classical feedback control' is all about and the intent of this book is to provide the reader with a fundamental insight into what makes closed loop systems tick and also to develop the tools that can be applied easily to both analysis and test.

2.1.1 The Simplest Form of Closed Loop System

Let us begin by analyzing the simplest form of closed loop system indicated by the block diagram of Figure 2.2. For the moment we will assume

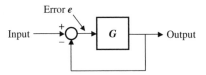

Figure 2.2 The simplest closed loop system example

that G is an algebraic constant, but remember that transfer functions typically contain dynamic terms involving the operator D. From the diagram we can write:

$$\text{output} = \text{e } G = (\text{input} - \text{output})G = (\text{input})G - (\text{output})G.$$

Gathering up the output terms on the left-hand side of the equation we obtain:

$$\text{output}(1 + G) = \text{input}(G).$$

Therefore:

$$\frac{\text{output}}{\text{input}} = \frac{G}{1 + G}.$$

The right-hand side of the above expression is referred to as the closed loop transfer function (CLTF) of the system depicted in Figure 2.2.

We can see from the above transfer function that if G is very large relative to 1.0 the value of the expression tends towards 1.0. Thus for the output to become close to, or equal to, the input command, the value of G must be large. For example, for $G = 100.0$ there will be a 1 % error between input and output. If there is an element K in the feedback path, as indicated by Figure 2.3, the response of the system becomes:

$$\frac{\text{output}}{\text{input}} = \frac{G}{1 + KG}.$$

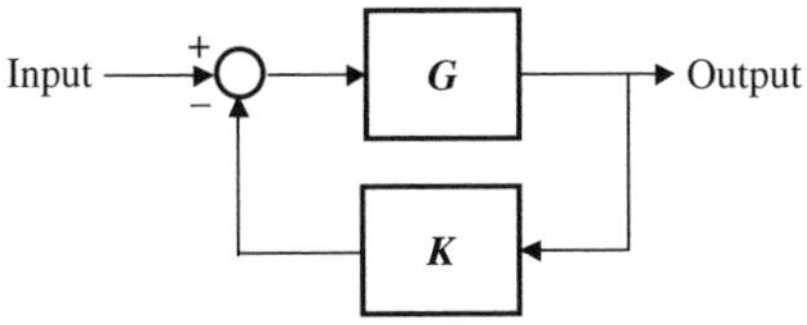

Figure 2.3 Simple system with feedback element

For a constant gain in the feedback path the response tends to $1/K$ as G becomes large. The rule for developing the relationship between the output and the input when there are elements in the feedback path is:

$$\frac{\text{output}}{\text{input}} = \frac{(\text{the product of all elements in the forward path})}{(1 + \text{the product of all elements around the loop})}.$$

Let us now move on to consider the dynamic behavior of closed loop systems and what causes them to become unstable.

2.2 The Concept of Stability

Stability, when referring to a closed loop control system, is a qualitative description of the performance of the system. Ideally, the output of a closed loop control system should respond quickly and precisely to changes in the input. In an attempt to improve the speed of response the designer can increase the gain of the controller; however, in doing so, the output response may begin to exhibit oscillatory behavior. This phenomenon is caused by time delays around the control loop and can lead to instability. When instability is reached, the output may oscillate continuously or the oscillations may continue to increase until the output reaches its maximum limit. To understand this effect let us return to the generic closed loop example given at the beginning of this chapter, with a few changes:

- insert a switch in the error signal line;
- assume that the input to the control loop is held constant;
- assume that the disturbance input is zero.

Now consider a sinusoidal input to the controller with the switch in the error path set open as indicated in the Figure 2.4. You can visualize from the diagram that if the feedback signal (measured output) is equal in magnitude and exactly 180 degrees out of phase with the originating signal, then the oscillations will continue if the switch is closed even with a fixed input to the control loop. In other words, the oscillation just continues going around and around at the same amplitude. When this condition occurs the system is said to have 'marginal stability'.

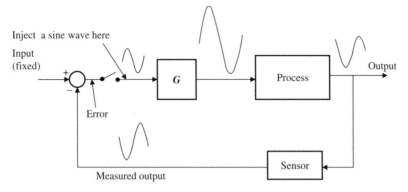

Figure 2.4 Illustration of the concept of stability

The rule for stability is:

In a closed loop system (with negative feedback) when the phase lag around the loop is 180 degrees (1/2 a cycle), the gain around the loop must be less than 1.0 for the system to be stable.

From this rule it follows that to determine the stability of a closed loop system we need to analyze the open loop behavior of the system. Let us look at what this means mathematically. Remember that the expression for the 'closed loop transfer function' (CLTF) is:

$$CLTF = \frac{(\text{forward path})}{(1 + \text{loop})}$$

where 'loop' is the product of all of the elements around the control loop.

Therefore if the denominator goes to zero, the CLTF becomes infinite. So the condition for marginal stability is satisfied when the dynamic elements around the loop combine to meet the condition:

$$1 + \text{Loop} = 0.$$

We can also rearrange the equation into the following form:

$$\text{Loop} = -1.0.$$

The right-hand side of this equation can be regarded as a vector quantity having a magnitude of 1.0 and a phase angle of 180 degrees, i.e.

$$|1.0|\angle - 180°$$

In the context of our diagrammatic example, this equation implies that the sine wave induced at the error signal switch is lagging by 180 degrees (one half of a cycle) by the time it gets all the way around the loop to the input summing junction. Also, the magnitude of the sine wave is 1.0 at this point (i.e. equal to the input stimulus). This fits well with the marginal stability definition stated above.

One final point that should be noted regarding the equation

$$1 + \text{loop} = 0.$$

In mathematical terms this is a differential equation referred to as the 'characteristic equation' of the control system. Since this equation contains all of the elements around the loop, its solution defines the transient response characteristics of the system. A better appreciation of this comment will develop as we proceed through this book.

2.3 Response Testing of Control Systems

The dynamic performance of a closed loop system can be expressed in terms of certain attributes of the system's 'response' to external stimuli. Such attributes include the ability to respond very quickly to changes in the input command versus a more docile reaction to changes. Fast response usually comes at a price. Typically, a sensitive, high gain control system may exhibit oscillatory tendencies. In some applications, a slower, more sluggish response with non-oscillatory behavior may be preferred. In some cases the response to a change in input command that results in an output which transiently overshoots the desired output may be totally unacceptable. For example, a control system used to control the position of a machine tool cannot tolerate overshoot since this would result in excessive material removal. In order to determine the qualitative behavior of closed loop systems we therefore need to establish procedures that can be used to determine numerically the dynamic characteristics of the system using either analytical or test methods. Both approaches are necessary in order to be able to verify consistency between theory and practice.

The term used to describe the dynamic behavior of a closed loop system is described as the 'response characteristic' of the system and can be determined as either:

- the transient response – i.e. the system reaction to a sudden change of input command; or
- the steady state response – i.e. the eventual steady state behavior of the system when subjected to a continuous time-varying input command such as a sinusoidal input at a specific frequency.

Figure 2.5 illustrates three types of response that can be used to assess the suitability of the dynamic performance of a closed loop system.

Response to a step input is used to evaluate the transient characteristics of a system. The sharp-edged feature of the step means that all frequencies are stimulated and the system response will contain the dynamic contributions from all of the roots of the characteristic equation:

$$1 + \text{loop} = 0.$$

While this type of stimulus is easy to apply to electrical and electronic systems it is more difficult to apply a pure step to systems with mechanical or fluid-mechanical components where rounded corners inevitably occur that will result in a less than pure result. The most important feature of step response is the ability to illustrate clearly the tendency of a system to overshoot the final condition which may be a critical aspect of the dynamic performance of some systems.

The ramp input is another form of input which provides an insight into both the transient and steady state responses of a system. This type of input is important in assessing a system's ability to follow a velocity command. An example would be target tracking where the target is moving at a specific velocity. The start of the ramp, being sharp-edged excites the transient roots of the system which eventually die away leaving the steady state response.

Frequency response (see Figure 2.5(c)) is perhaps the most common method for evaluating the dynamic performance characteristics of a

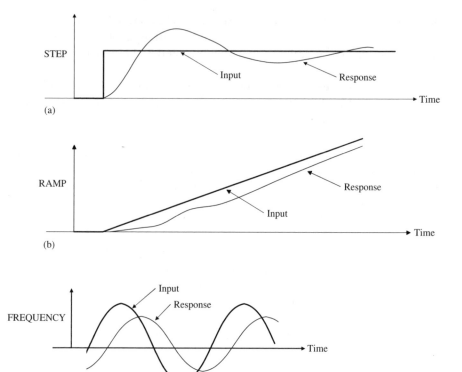

Figure 2.5 (a) Step response; (b) ramp response; (c) frequency response

system. The input command is varied sinusoidally and the output response in terms of its relative amplitude and phase shift identified. Analysis and/or testing can focus on a specific number of frequencies in the range of interest. For (mostly) linear systems sinusoidal inputs generate (mostly) sinusoidal outputs with attenuation (or magnification) and a time delay (also referred to as a phase shift or phase lag expressed in degrees).

If we go back to our generic closed loop system let us consider varying the command signal sinusoidally as indicated in Figure 2.6. Typically, we observe that at frequencies below a certain value, the output follows the command closely with little attenuation or phase shift. Above this frequency, the system begins to struggle to keep up and the output becomes more and more attenuated and

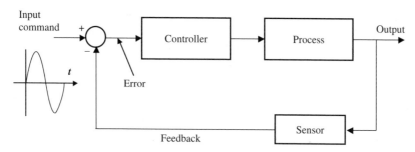

Figure 2.6 Closed loop system with sinusoidal input command

delayed in time (see Figure 2.7). Frequency response is expressed as the ratio between the output amplitude and the input amplitude plotted against frequency together with a plot of phase lag against frequency.

A word of caution is worth mentioning here regarding the use of transducers and instrumentation in a dynamic response test arrangement. Great care must be taken to ensure that the transducers selected to measure the various system variables have response characteristics that are ideally a couple of orders of magnitude faster than the system under test. The test setup is equally critical particularly when measuring fluid pressures and flows. Flexible hoses and pressure gauges can significantly affect the measured response by virtue of the capacitance effect caused by the change in volume that occurs during pressure changes. If the necessary steps are not taken to avoid these problems then the test results may be seriously compromised.

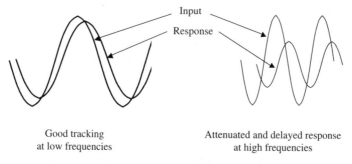

Good tracking   Attenuated and delayed response
at low frequencies   at high frequencies

Figure 2.7 Typical frequency response

Before proceeding further in describing methods and procedures for analyzing the response of control systems, we need to address a number of key aspects of system behavior and how graphical methods have evolved that simplify the analytical process.

2.4 The Integration Process

A good place to start is to discuss the process of 'integration' as it applies to the dynamic behavior of control elements and systems. Integration is probably the most important function associated with control systems. Common examples of integration include:

$$\text{car heading} = \int (\text{steering wheel angle}) \mathrm{d}t.$$

This means that if the steering wheel angle is in neutral (i.e. zero) the car heading will not change. However, if the wheel is deviated to one side the car heading will continue to change. The bigger the deviation, the faster the rate of change of heading.

$$\text{fluid volume} = \int (\text{control valve position}) \mathrm{d}t;$$

here control valve position determines fluid flow rate (assuming a pressure source upstream of the valve) and the output volume passed through the valve is the integral of flow rate and hence valve position. The more open the control valve the faster the volume will increase. Let's see how this is expressed in block diagram form. Using the D notation developed previously we can express the above equations by the block diagrams of Figure 2.8.

Listed below are two key frequency response characteristics of the integration process that must be understood. These characteristics will form an important part of the analytical processes that will follow.

- At all frequencies there is a 90 degree phase lag (a quarter of a cycle) introduced by the integration process. (To understand this statement, try to plot the car heading that results from a sinusoidal oscillation of the steering wheel angle and you will clearly see this phase angle shift effect.)

- When the frequency of the input is doubled, the amplitude at the output of the integrator is halved. Likewise, when the frequency is halved the output amplitude is doubled. (Intuitively this makes sense. Consider steering your car. As you oscillate the steering wheel angle very slowly from left to right the car will make large oscillations in heading. But when you waggle the steering wheel quickly over the same amplitude the car will wobble a little but the basic heading will not change significantly.)

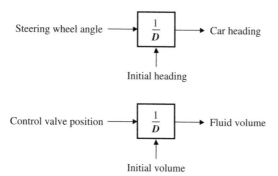

Figure 2.8 Integration examples in block diatgram form

To illustrate these principles using control terminology, let us consider an example based on the control valve integration referred to above. Referring to the block diagram of Figure 2.9, we have a control valve (shown as a gate valve with a linear position x_V) controlling the flow

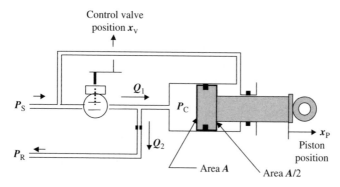

Figure 2.9 Hydraulic actuator block diagram

into a half-area actuator. Supply pressure P_S is applied to the control valve and also to the half-area side of the servo-actuator piston. A fixed orifice connects the control pressure side of the control valve P_C to the hydraulic return pressure P_R, which we can assume to be nominally zero. This arrangement allows movement of the control valve to cause movement of the piston in either a positive or negative direction.

Assuming there is no load on the actuator, the control pressure P_C must always be equal to $P_S/2$, and the flow through the control valve will determine the velocity of the piston. The flow through the fixed orifice will therefore be constant since P_C and P_R are constant.

To evaluate the dynamic behavior of the actuator (again assuming there is no external load on the actuator and also that inertial and friction forces are negligible) we can write the equations of motion:

control valve flow $Q_1 = K x_V$, where K is the valve 'flow gain' in fluid volume per second per unit displacement of x_V,
fixed orifice flow Q_2 will remain constant for a constant P_C,
piston velocity $= (Q_1 - Q_2)/A$, i.e $D x_P = (K x_V - Q_2)/A$.

While this example is a good representation of an integration process, the reader should be aware that when carrying out this test in practice, keeping the piston in a constant mid-stroke position is difficult because only the slightest imbalance between Q_1 and Q_2 would result in the piston drifting slowly to one extreme.

We can continue our analytical exercise by expressing these equations in block diagram form as indicated in Figure 2.10. Since Q_2 is a constant, it can be ignored for the purpose of understanding the dynamics of this system. Thus we have normalized the system about the steady state point which allows us to examine perturbations of x_V and x_P . Consolidating the above diagram to get a single transfer function results in the simple block diagram of Figure 2.11.

Let us check the units for the final transfer function.

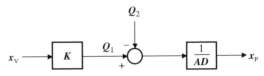

Figure 2.10 Hydraulic value and actuator block diagram

Figure 2.11 Hydraulic valve and actuator consolidated block diagram

US/Imperial units:

$$(\text{in.}) \left(\frac{\text{in.}^3}{(\text{sec})(\text{in.})} \right) \left(\frac{1}{\text{in.}^2} \right) (\text{sec}) = X_\text{P} = (\text{in.}) \quad \checkmark$$

SI units:

$$(\text{m}) \left(\frac{\text{m}^3}{(\text{s})(\text{m})} \right) \left(\frac{1}{\text{m}^2} \right) (\text{s}) = X_\text{P} = (\text{m}) \qquad \checkmark$$

Here K/A is termed the 'gain' of the integrator and has units of $(\text{seconds})^{-1}$ which is the same as radians per second, i.e. frequency. The gain of the integrator is the frequency at which the amplitude ratio between the input to the integrator and the output response is unity. In this example, a quarter of an inch oscillation of x_V at a frequency of K/A radians per second will produce a quarter of an inch oscillation at the piston x_P. (Note also that the oscillation of x_P will be displaced in time by a quarter of a cycle (90 degrees) from the input x_V per the first integration characteristic listed above.)

Substituting values of $K = 10.0$ and $A = 1.0$ in the above example allows us to generate a plot of amplitude ratio against frequency as shown in Figure 2.12. Here the integrator 'gain' $K/A = 10.0 \text{ s}^{-1} \ (\text{sec}^{-1})$ and therefore the amplitude ratio is equal to unity at a frequency of 10 radians per second as shown in the figure.

The characteristic of amplitude doubling and halving with frequency halving and doubling is also shown. To better accommodate this feature a convention is used that converts amplitude ratio into a logarithmic equivalent called 'decibels' expressed as dB. The adopted convention is:

$$\text{decibels (dB)} = 20 \log_{10} (\text{amplitude ratio}).$$

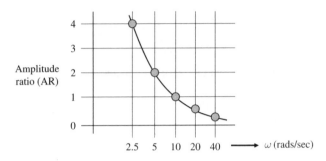

Figure 2.12 Amplitude ratio vs frequency for the valve and actuator

Therefore:

$$AR = 1.0 = 0\,\text{dB}$$

$$AR = 2.0 = +6.02\ \text{dB}$$

$$AR = 0.5 = -6.02\ \text{dB}$$

$$AR = 10 = +20\ \text{dB}$$

and so on. When amplitude ratio is converted to decibels it is called 'gain'. Frequency is also plotted on logarithmic scales using either ω radians per second or f hertz.

Figure 2.13 is a repeat of Figure 2.12 using the gain convention described above. Note that we now have a constant slope gain line.

An important observation from both Figures 2.12 and 2.13 is that as the frequency of oscillation tends to zero, the gain tends to infinity. This

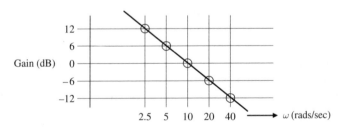

Figure 2.13 Gain vs frequency for the valve and actuator

feature is important in the design of control systems since it means that the presence of an integration process in a closed loop control system will result in zero error in steady state (i.e. at zero frequency). Other conventions are used in frequency response jargon:

- doubling the frequency increases frequency by an 'octave';
- a factor of 10 change in frequency is called a 'decade'.

Thus the slope of the line in Figure 2.13 is:

$$-6\,\mathrm{dB} \text{ per octave or } -20\,\mathrm{dB} \text{ per decade.}$$

Let us summarize the effects of integrators in closed loop control systems:

- Since an integrator has infinite gain at zero frequency, a system with a single integrator in the loop will exhibit zero steady state error. (Note that the integrator may be designed in as part of the controller function or it may be an inherent feature of the process being controlled.)
- It can be shown that a system with two integrators in the loop will have zero error for inputs moving at constant velocity.
- Similarly, a system with three integrators will have zero error with a constant acceleration input command.

With each integrator in the loop comes a quarter cycle delay at all frequencies. Two integrators would incur a half cycle delay around the loop equivalent to 180 degrees of phase lag. Therefore, even though the integration process has major benefits to control systems by eliminating steady state and even dynamic errors, there are attendant stability issues that must be dealt with in the control system design process.

2.5 Hydraulic Servo-actuator Example

Let us now consider a more sophisticated example. Figure 2.14 shows a schematic of a typical power control unit (PCU) for an aircraft flight control surface. The input end of the summation link is connected via cables and pulleys to the pilot's control column. This hydraulic position servo actuator allows the pilot to overcome the large hinge moments that occur during high speed flight.

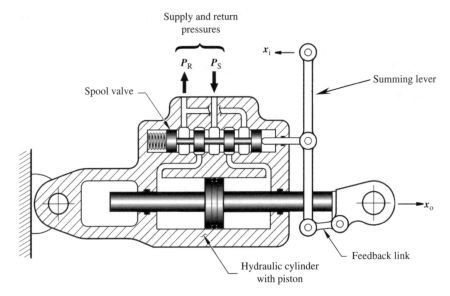

Figure 2.14 Aircraft flight control PCU schematic

Movement of the mechanical input (x_i) moves the spool valve to the left causing hydraulic fluid to flow from the supply pressure source P_S into the piston chamber to the left of the piston. At the same time the right side of the piston is opened to the return line to P_R. The resulting piston movement (x_o) to the right continues until the summing lever has returned the spool valve to the neutral or null position. Figure 2.15 is a block diagram representation of the actuator. The linkage gearing G will be equal to 0.5 for a summing lever where the error pivot is equidistant from the input and output pivot points.

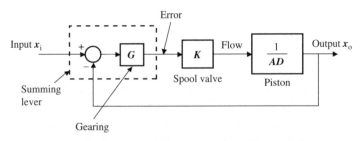

Figure 2.15 Actuator block diagram

It should be noted that this approach to hydraulic actuator dynamic analysis is greatly simplified in the interest of developing an easy-to-understand introduction into control system analysis. Specifically the inertia and external forces on the actuator have been neglected. In spite of this radical simplification, the dynamic model developed here is quite adequate for use in aircraft dynamic analysis provided that the external forces represent less than about 25% of the available pressure drop because the spool valve flow gain is a function of the square root of the pressure drop across it. Thus a 25% reduction in available pressure drop manifests itself as only an 11% reduction in flow gain.

From Figure 2.15 we can now develop the closed loop transfer function (CLTF) using the methods presented earlier.

$$\text{CLTF} = \frac{x_o}{x_i} = \frac{\text{forward path}}{1 + \text{loop}} = \frac{\left(\dfrac{GK}{AD}\right)}{\left(1 + \dfrac{GK}{AD}\right)}$$

i.e.

$$\frac{x_o}{x_i} = \frac{1}{(1 + TD)} \quad \text{where} \quad T = \frac{A}{GK}.$$

Note the similarity of this example to the previous integrator example. In this case the integrator has a position feedback around it which causes the spool valve to return to the null position following a change in the input command.

For the closed loop transfer function developed above, we can make these following observations:

- The transfer function of the above form is called a 'first order lag' since the highest order of the D operator is 1.
- In the term $T = A/GK$, T is called the 'time constant' and has units of time (check units: $\mathrm{m^2/\left(m^3/s\right)/m = s}$ or $\mathrm{in.^2/\left(in.^3/sec\right)/in. = sec}$).
- When the dynamic term TD is large relative to unity (i.e. at high frequencies) the response x_o/x_i tends to 1/TD which looks like an integrator.
- Similarly when things are changing slowly the term TD becomes very small relative to unity hence the closed loop response x_o/x_i tends to 1.

2.6 Calculating Frequency Response

Frequency response is expressed as the ratio between the output vector and the input vector as a function of $j\omega$. Mathematically we can say:

$$\frac{x_o}{x_i}(j\omega) = \left|\frac{x_o}{x_i}\right| \angle \left(\frac{x_o}{x_i}\right)$$

Here $\left|\dfrac{x_o}{x_i}\right|$ is the modulus or amplitude ratio of the response and $\angle\left(\dfrac{x_o}{x_i}\right)$ is the phase angle between the input and output vectors.

In calculating the frequency response of a transfer function we need to develop simple methods of determining the amplitude ratio and phase angle and then plotting the results against frequency. To calculate these terms we need to revisit the complex sinusoid described in Chapter 1. Let us consider the input to the system to be a sinusoid represented by a vector rotating at ω radians per second as shown in Figure 2.16.

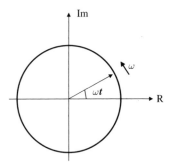

Figure 2.16 Rotating vector in the complex plane

This vector can be expressed in the form of the complex sinusoid described in Chapter 1, i.e. $e^{j\omega t}$, Let us use the closed loop transfer function developed above for the flight control actuator as the system for which we want to develop a frequency response analysis. This is represented by the block diagram in Figure 2.17.

Assuming an input to the system x_i of the form $xe^{j\omega t}$ we can develop the differential equation relating the input forcing function

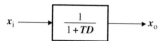

Figure 2.17   PCU closed loop transfer function

and the output response from the closed loop transfer function as follows:

$$xe^{j\omega t} = x_o + TDx_o.$$

Let us assume that the solution for x_o in the above equation is $ye^{j\omega t}$. This assumes that the output is a sinusoid of the same frequency as the input forcing function which is reasonable.

Substituting this solution into the last equation we have:

$$xe^{j\omega t} = ye^{j\omega t} + TDye^{j\omega t} = Ye^{j\omega t} + T(j\omega)ye^{j\omega t}.$$

Rationalizing we obtain:

$$\frac{ye^{j\omega t}}{xe^{j\omega t}} = \frac{x_o}{x_i}(j\omega) = \frac{1}{[1+(j\omega)T]}.$$

All we have done is to replace the operator D by the term $j\omega$.

The equation states that the response of this system to a sinusoidal input is simply a complex number expressed as a function of the input frequency ω. The left-hand side of the equation is the ratio of two vectors at any given frequency and can be represented by two specific input and output vectors in the complex plane as indicated by Figure 2.18.

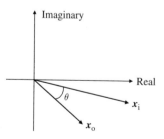

Figure 2.18   Input and output vectors at a specific frequency

The 'amplitude ratio' of the response is the ratio of the lengths (moduli) of the vectors and the phase shift is the phase angle difference between the two vectors, i.e.

$$\text{amplitude ratio (AR)} = \left| \frac{x_o}{x_i} \right|$$

$$\text{phase angle}(\theta) = \angle x_i - \angle x_o.$$

To summarize, the frequency response can be calculated by following the simple rules below:

1. Substitute $D = (j\omega)$ in the transfer function.
2. Plug in values of ω covering the frequency range of interest.
3. Gather up the real and imaginary terms at each frequency.

If we assume that the forcing function vector always lies along the positive real axis in the complex plane as indicated by the diagram of Figure 2.19 we can say:

$$\text{Amplitude ratio} = \sqrt{(R^2 + I_m^2)}$$

$$\text{Phase shift}(\theta) = \tan^{-1} \left(\frac{I_m}{R} \right).$$

Here R and I_m are the real and imaginary components of the output vector respectively.

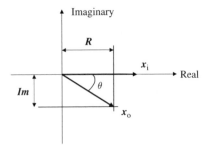

Figure 2.19 Normalized response vectors in the complex plane

2.6.1 Frequency Response of a First-order Lag

Having developed the basic rules for calculating the amplitude ratio (or gain) and phase angle of transfer functions, let us now calculate the frequency response of the generic first-order lag defined as follows:

$$\frac{1}{(1 + (j\omega)\,T)}$$

Based on the rules developed above we can say:

$$AR = \frac{1}{\left[1 + (\omega T)^2\right]^{\frac{1}{2}}} \quad \text{and} \quad \theta = -\tan^{-1}(\omega T).$$

Table 2.1 shows the amplitude ratio (AR), gain and phase angle values for frequencies on either side of $\omega = 1/T$ radians per second.

We can now plot the results for gain and phase angle against frequency as shown in Figure 2.20. Note that the gain curve is 'asymptotic' to zero dB for low frequencies and asymptotic to $1/TD$ (an integrator) at the higher frequencies which is a straight line with a slope of -6.0 dB/octave.

These asymptotes meet at $\omega = 1/T$ radians per second. This frequency is referred to as the 'bandwidth' or 'break frequency' of the first-order lag.

The plotting convention used in the above figure shows the phase lag increasing upwards on the graph. This is opposite to the traditional convention used by most textbooks since it can be argued that phase

Table 2.1 First-order lag frequency response

ω	ωT	$(\omega T)^2$	$1 + (\omega T)^2$	AR	Gain (dB)	$\theta°$
$1/8T$	0.125	0.016	1.016	0.997	-0.1	-7
$1/4T$	0.25	0.063	1.063	0.970	-0.3	-14
$1/2T$	0.5	0.25	1.25	0.894	-1.0	-27
$1/T$	1.0	1.0	2.0	0.707	-3.0	-45
$2/T$	2.0	4.0	5.0	0.448	-7.0	-63
$4/T$	4.0	16.0	17.0	0.243	-12.3	-76
$8/T$	8.0	64.0	65.0	0.124	-18.1	-83

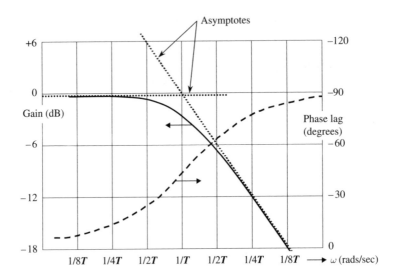

Figure 2.20 Frequency response of a first-order lag

lag is a negative quantity and therefore should increase downwards. The convention used here is, in the author's opinion, easier to view because the gain and phase curves go in opposite directions. Using the more traditional convention can be confusing when the gain and phase lines have similar shapes often lying on top of each other. In any case these are simply conventions and the reader should feel free to adopt either method.

In order to expedite the process of control system analysis which involves the generation of frequency response plots for the various elements around the loop it will save time to commit to memory some key numbers so that these plots can be quickly sketched without having to resort to calculation.

Some useful numbers to remember are:

- At the break frequency the gain is −3 dB and the phase lag is 45 degrees.
- At half the break frequency the gain is −1 dB and the phase lag is 27 degrees.
- At twice the break frequency the gain is −7 dB and the phase lag is 63 degrees.
- At higher and lower frequencies the gain approximates the asymptotes and the phase lag tends to 90 degrees and 0 degrees respectively.

From these few 'magic numbers' it is easy to generate frequency response plots of first-order lags quickly as part of the design and analysis process.

To obtain the response of two or more terms in series the gains and phase angles are simply added together as we will see in up-coming examples. When the above type of graph is used to plot the 'open loop' characteristics of a system, it is referred to as a 'Bode diagram'. More about this later.

2.6.2 Frequency Response of a Second-order System

To reinforce the frequency response analysis process described above we shall now apply it to the standard form of second-order system developed in Chapter 1 as part of the spring–mass system example. The transfer function relating the input and output displacements is:

$$\frac{x_o}{x_i} = \frac{1}{\left(\dfrac{D^2}{\omega_n^2} + \dfrac{2\zeta D}{\omega_n} + 1\right)}$$

where ω_n is the undamped natural frequency and ζ is the damping ratio. As before, we simply put $D = j\omega$ in the above transfer function, plug in values for ω gathering up the real and imaginary parts. Table 2.2 shows the frequency response values for the spring–mass system for several frequencies either side of the undamped natural frequency ω_n using a value for the damping ratio of $\zeta = 0.2$.

Table 2.2 Second-order system frequency response Frequency response of $\dfrac{1}{((j\omega)^2/\omega_n^2 + 2\zeta(j\omega)/\omega_n + 1)}$ for $\zeta = 0.2$

ω	$(j\omega)^2/\omega_n^2$	$2\zeta(j\omega)/\omega_n$	R	Im	AR	Gain (dB)	$\theta°$
$\omega_n/8$	$-1/64$	$0.05j$	0.984	0.05	1.015	0.13	-3
$\omega_n/4$	$-1/16$	$0.1j$	0.963	0.1	1.06	0.51	-6
$\omega_n/2$	$-1/4$	$0.2j$	0.750	0.2	1.29	2.21	-15
ω_n	-1.0	$0.4j$	0	0.4	2.5	7.96	-90
$2\omega_n$	-4.0	$0.8j$	-3.0	0.8	0.33	-9.63	-165
$4\omega_n$	-16.0	$1.6j$	-15.0	1.6	0.067	-23.6	-174
$8\omega_n$	-64.0	$3.2j$	-63.0	3.2	0.016	-35.9	-177

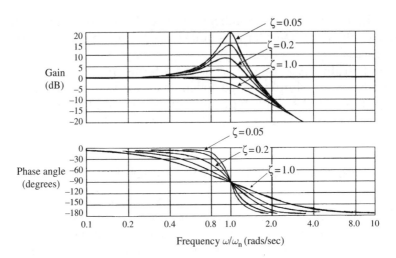

Figure 2.21   Second-order system standard response curves

The graph of Figure 2.21 shows standard frequency response plots for second-order systems for several values of damping ratio. Standard curves such as these are available to the control system designer to simplify the analysis task associated with the development of composite gain and phase plots through the addition of the various elements around the loop.

From Figure 2.21 we can make a number of observations.

1. The damping ratio determines the degree of magnification seen at the output. Specifically, for $\zeta = 0.05$, magnification of the input signal is approximately 10 times (20 dB). For damping ratios below 0.2, the peak magnification can be approximated to $1/2\zeta$.
2. As the damping ratio is increased the peak magnification reduces and the frequency corresponding to maximum response continues to reduce to a value somewhat lower than ω_n.
3. The phase lag for all damping ratios increases to 90 degrees of lag at the undamped natural frequency ω_n continuing on towards a maximum of 180 degrees of lag at the higher frequencies.
4. The lower damping ratios exhibit a more sudden transition in phase lag from frequencies below ω_n to the frequencies above ω_n than the higher damped systems.

5. For damping ratios of 1.0 or greater there is no magnification of the input signal and the second-order system can be represented by two first-order lags in series, i.e. the roots become real.

This is all we need to know for deriving the frequency response characteristics since a third-order system is made up of either three first-order elements or one first-order element and a second-order element and so on for higher-order systems.

2.7 Aircraft Flight Control System Example

We will now use a simplified aircraft flight control system example to apply what we have learned through the analysis of the open loop frequency response. From this we will establish the stability margins of the system which represent the qualitative dynamic performance of the system.

Figure 2.22 is a schematic of an aircraft pitch attitude control system example wherein the flight control computer determines and controls the required aircraft pitch attitude. The command signal is compared with the measured pitch attitude, determined via the inertial reference system, to establish an error signal which in turn drives a servo actuator

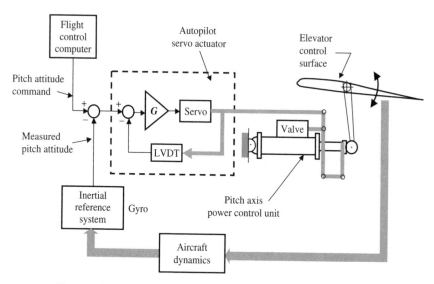

Figure 2.22 Aircraft pitch attitude control system

whose output positions the input linkage on the pitch axis elevator power control unit (PCU). The PCU can be regarded as a dynamic element with essentially the same form of response characteristics as the actuator analyzed in Section 2.5. The PCU output drives the elevator control surface causing the aircraft to respond in pitch.

2.7.1 Control System Assumptions

We now need to establish numerical values for the functional constants and dynamics of the various elements in the control loop so that we can apply our analysis techniques based on the methods outlined earlier.

With regard to the PCU, based on what was mentioned earlier, the effect of aerodynamic hinge moments which the actuator must overcome in order to move the control surface can be ignored provided that the force reflected back to the actuator is equivalent to not more than 25 % of the maximum available pressure drop. This condition will be assumed to be valid in the control element values developed below. For the purpose of this design study the following assumptions and design parameters will be adopted.

- The avionics, including the inertial reference system (gyro) are fast relative to the other elements in the control loop and therefore may be neglected.
- The autopilot servo actuator has a first-order lag of 0.02 seconds and an output displacement of 0.4 inches per degree of pitch attitude error.
- The PCU input-to-elevator angle ratio is 10 degrees per inch and the PCU dynamics are equivalent to a first-order lag of 0.1 seconds.
- Aircraft dynamics relating elevator angle to aircraft pitch attitude can be represented by the transfer function of Figure 2.23.

We can make an observation from this transfer function regarding how the aircraft will respond to changes in elevator angle. The presence of the D term in the denominator (which is the definition of an

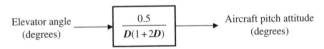

Figure 2.23 Aircraft dynamics transfer function

integrator) implies that elevator deviations from steady state will result in a rate of change of pitch attitude. Also the bigger the elevator deviation, the faster the pitch attitude will change. The second term is a first-order lag with a time constant of 0.2 seconds which adds an additional delay into the aircraft response process. It should be realized here that this simple transfer function is only an approximation and would not be valid for large excursions in pitch.

2.7.2 Open Loop Analysis

We can now construct a control system block diagram that incorporates all of the above numerical assumptions for subsequent analysis. This diagram is presented as Figure 2.24. In order to determine the stability attributes of this control system we need to evaluate the open loop performance of the system to assess how the open loop response relates to the stability criterion established earlier.

From the block diagram described by Figure 2.24, the open loop transfer function (OLTF) is defined by multiplying together all of the 'blocks' from the error signal all the way around the control loop. The result is:

$$\text{OLTF} = \frac{(0.4)\,(10.0)\,(0.5)}{D\,(1+0.02D)\,(1+0.1D)\,(1+2.0D)}$$
$$= \frac{2.0}{D\,(1+0.02D)\,(1+0.1D)\,(1+2.0D)}.$$

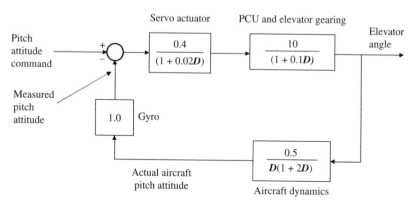

Figure 2.24 Pitch attitude control system block diagram

From here we can develop frequency response plots for each of the terms in the OLTF. Beginning with the 'gain' aspect of the OLTF and adopting the dB convention, the product of all of the terms in the OLTF is simply the graphical addition of each of the terms in the OLTF as indicated in Figure 2.25.

Note the integrator term, $2/D$. This term has a slope of -6 dB per octave and crosses the 0 dB line at a frequency of 2 radians per second. The term $1/(1 + 2.0D)$ is a first-order lag with a time constant of 2.0 seconds. The asymptotes of a first-order lag gain are a constant 0 dB at low frequencies up to the break frequency which is $1/(2.0)$ or 0.5 radians per second. Beyond the break frequency the gain slope becomes -6.0 dB per octave.

All of the transfer function terms can be plotted separately in this way and the combined gain response of the open loop system obtained by adding each of the terms together graphically. The result shows the gain graph beginning at the far left attenuating with a slope of -6.0 dB per octave. This slope continues up to 0.5 radians per second which is the break frequency of the aircraft 2.0 seconds first-order lag term. The gain slope now changes to -12.0 dB per octave and remains constant

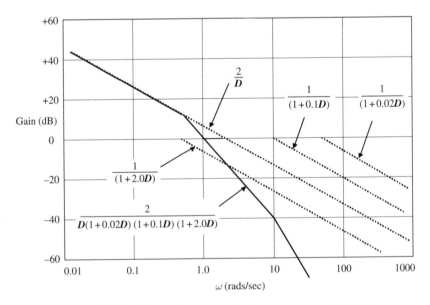

Figure 2.25 Open loop gain frequency response

at this value up to the break frequency of the PCU actuator which is 10 radians per second. Above this frequency, the gain slope changes to -18.0 dB per octave. The servo actuator, with its break frequency at $1/(0.2)$ or 50 radians per second, does not contribute significantly to the final gain curve.

We can also construct a phase angle versus frequency plot in the same way. This is shown as Figure 2.26. Once again we simply add the phase contributions of each transfer function to obtain a combined total. Here the phase lag begins with a lag of 90 degrees due to the integrator and rapidly climbs to more than 180 degrees at about 2 radians per second and above.

In order to assess the stability of this control system we must now combine the open loop gain and open loop phase plots on a single chart. This is shown as Figure 2.27 and is referred to as a *'Bode diagram'*. Our phase lag plot convention discussed earlier is particularly useful in Bode diagrams since we can arrange the 180 degree phase lag line to coincide with the 0 dB line. Thus the stability margins are easy to see.

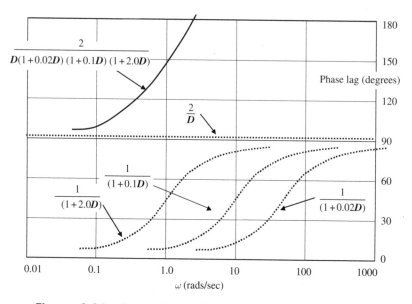

Figure 2.26 Open loop phase frequency response

Closing the Loop

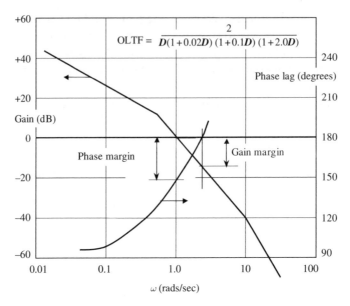

$$\text{OLTF} = \frac{2}{D(1+0.02D)(1+0.1D)(1+2.0D)}$$

Figure 2.27   The Bode diagram for the system

From the above Bode diagram the stability margins for the system are defined as follows.

Gain margin: The gain increase that would result in the gain curve crossing the 0 dB line at the frequency corresponding to 180 degrees of open loop phase lag.

Phase margin: The additional phase lag that would result in 180 degrees of open loop phase lag at the frequency corresponding to 0 dB of open loop gain.

In our example, the gain and phase margins shown on Figure 2.27 are approximately 14 dB and 30 degrees respectively.

In practical terms this means that if the open loop gain was increased by 14 dB, which is a factor of about five, the system would be unstable. Similarly if an additional phase lag of 30 degrees were to be present at a frequency of 1 radian per second the system would be unstable. As a general guide, good design practice is to aim for a minimum of 6 dB (i.e. a factor of two) gain margin and a phase margin of 45 degrees. In our case the gain margin is very good at about 15 dB but the phase margin of 30 degrees is not sufficient. This indicates that the system is 'phase sensitive'.

2.7.3 Closed Loop Performance

The next step is to understand what all this open loop analysis means in terms of system behavior when operating as a closed loop system. Going back to the system block diagram of Figure 2.24 we can generate the closed loop transfer function (CLTF) for the complete system using the rule:

$$\frac{\text{output}}{\text{input}} = \frac{(\text{forward path})}{(1 + \text{loop})}.$$

In this case the aircraft response in pitch attitude θ to the required pitch attitude command θ_C is:

$$\frac{\theta}{\theta_C} = \frac{\left[\dfrac{2.0}{D\,(1+2.0D)\,(1+0.1D)\,(1+0.02D)}\right]}{1 + \left[\dfrac{2.0}{D\,(1+2.0D)\,(1+0.1D)\,(1+0.02D)}\right]}$$

$$= \frac{1}{[1 + (0.5D)\,(1+2.0D)\,(1+0.1D)\,(1+0.02D)]}$$

Multiplying out the denominator in the above transfer function will yield a fourth-order expression which can be factorized into the product of first-order and second-order terms. In this case the closed loop transfer function approximates to one second-order and two first-order terms as follows:

$$\frac{\theta}{\theta_C} = \frac{1}{(D^2 + 0.4D + 1)\,(1+0.1D)\,(1+0.02D)}.$$

The second-order term has a natural frequency of 1.0 radian per second with a damping ratio of 0.2. Figure 2.28 shows the closed loop frequency response of the system.

The resonance at 1.0 radian per second is quite marked indicating that the introduction of additional lag terms into the loop may result in more oscillatory behavior and eventually instability.

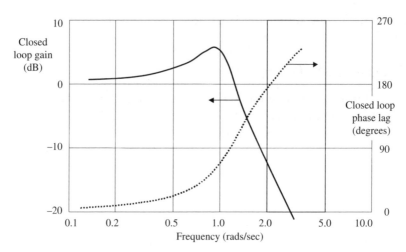

Figure 2.28 Closed loop frequency response

This method of transposing open loop frequency response to closed loop response is tedious when high orders of D are involved. There are, however, much more straightforward graphical tools that are readily available to provide the analyst with an immediate visualization of closed loop behavior directly from the open loop response. These alternative graphical methods are developed in the following section.

2.8 Alternative Graphical Methods for Response Analysis

In the last section we developed the concept of the Bode diagram which presents the open loop gain and phase characteristics of a closed loop control system from which we can readily determine gain and phase stability margins. Here we will address alternative methods of describing control system performance from which we can more easily transpose what we have learned from the open loop characteristics into how that system behaves as a closed loop system.

2.8.1 The Nyquist Diagram

The Nyquist diagram is one alternative to the Bode diagram. The Nyquist diagram uses the polar coordinates of the real and imaginary contributions of the open loop transfer function (OLTF) to illustrate the open loop frequency response characteristics of a system. With this graphical method, instead of having two curves plotted against frequency, one for gain and a second for phase angle, a single curve incorporating both gain and phase is presented as a single locus of the output vector with frequency being shown as specific points on the curve.

Another feature of the traditional Nyquist plot is that a polar coordinate representation is used with the amplitude ratio defined as the length of the output vector and phase angle represented by the angle of the output vector relative to the positive real axis.

Figure 2.29 shows the Nyquist plot of a simple first-order lag transfer function arbitrarily defined as

$$\frac{3.6}{(1 + (j\omega)\, T)}$$

to illustrate the concept. Note that the input command to the system is represented by a vector of unit magnitude in line with the real axis of the diagram (this is typically assumed but not shown in Nyquist plots). The curve depicts the locus of the response vector as frequency is increased from zero to infinity.

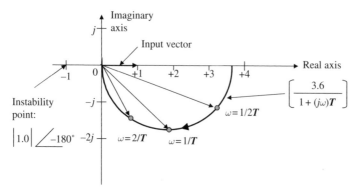

Figure 2.29 Nyquist diagram of a first-order lag

Specific frequencies along the curve are identified below for comparison with Table 2.1:

$$\omega = 1/2T \quad AR = 0.894 \quad \theta = -27°$$

$$\omega = 1/T \quad AR = 0.707 \quad \theta = -45°$$
$$\omega = 2/T \quad AR = 0.448 \quad \theta = -63°.$$

Note that the above amplitude ratio (AR) values are relative to the nominal zero frequency response of 3.6 (for an input of 1.0). Thus for an AR = 0.894, the output vector length is (3.6)(0.894) = 3.218. Note also that the locus of the response stays well clear of the instability point. In fact since the maximum phase angle for a first-order lag is only −90 degrees a system with only one first-order transfer function around the loop can never be unstable.

Let us now see how a number of different transfer functions look on the Nyquist diagram. Figure 2.30 shows Nyquist plots for typical systems having zero, one and two integrators in the open loop transfer function. These systems are referred to as *class '0'*, *class '1'* and *class '2'* systems respectively. The Nyquist curves on the left show unstable systems and the right-hand side curves show stable response plots.

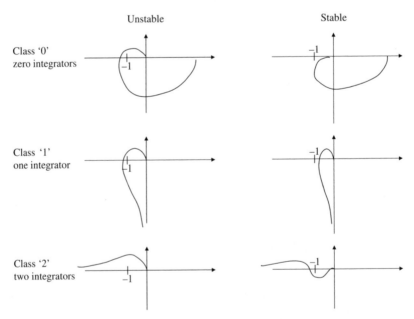

Figure 2.30 General examples of Nyquist plots

The *class* '0' system begins in phase with the input command at zero frequency and as the frequency increases the phase lag increases and the amplitude ratio (length of the response vector) decreases towards zero at infinite frequency. In the case of the unstable example, the phase lag passes 180 degrees while the amplitude ratio (response vector length) is still greater than 1.0.

The *class* '1' system begins with a phase lag of 90 degrees due to the integrator, and again the phase lag increases and the amplitude ratio decreases as before. For stability the locus must pass inside the (−180 degrees, −1.0) point on the graph.

The *class* '2' system begins with a phase lag of 180 degrees because of the presence of two integrators in the loop. For this system to be stable some means of 'reducing' the phase lag in the region of the (−180 degrees, −1.0) point must be introduced. This is achieved using performance compensation methods that will be described in the next chapter.

In order to reinforce the concept of the Nyquist diagram, Figure 2.31 shows Nyquist plots for three specific open loop transfer functions (OLTFs) including the aircraft control system example analyzed in

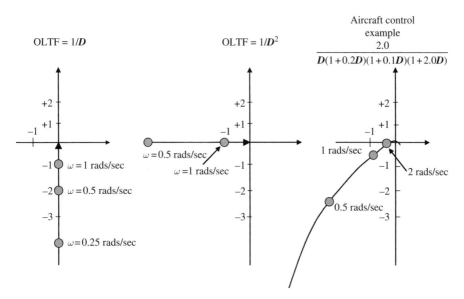

Figure 2.31 Specific examples of Nyquist plots

Section 2.7. The locus for $1/D$ originates at minus infinity on the y-axis continuing straight up towards the origin as the frequency increases maintaining a constant phase lag of 90 degrees. The second locus of $1/D^2$ comes from minus infinity on the left towards the origin maintaining a constant phase lag of 180 degrees. In doing so it passes through the instability point (-180 degrees, -1) at a frequency of one radian per second. This system would oscillate constantly at a frequency of 1 radian per second and by definition is 'marginally unstable'. The third locus is the aircraft control system example. The locus passes just inside the instability point indicating that the system is stable but the close proximity to that point suggests that the closed loop behavior may be somewhat oscillatory.

From the Nyquist diagram, we can also identify the specific gain and phase margins as was done using the Bode diagram. Figure 2.32 shows the Nyquist diagram for the aircraft control example magnified around the instability point indicating the stability margins. Referring to Figure 2.32, the gain margin in dBs is calculated by taking the reciprocal of the distance 'x' on the graph which is the distance from the origin to the point where the locus crosses the horizontal axis. In this case $1/X$ is approximately $1/0.2$. This means that the gain can be increased by a factor of five before instability occurs. This is equivalent to $+14$ dB as indicated on the original Bode diagram. The phase margin is defined as how much additional phase lag is needed to cause the locus to pass through the instability point at the frequency where the response vector

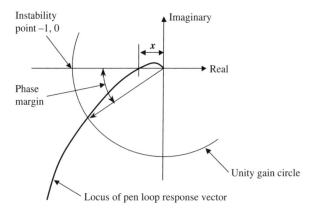

Figure 2.32 Nyquist diagram showing gain and phase margins

magnitude is 1.0. This can be seen directly from the graph at about 30 degrees which is again in agreement with the earlier Bode diagram developed as Figure 2.27.

2.8.2 Deriving Closed Loop Response from Nyquist Diagrams

While the stability attributes of a closed loop control system are established through the examination of open loop system characteristics, it is the performance of the control system when operating as a closed loop system that is also of critical interest to the control system designer and analyst. Slogging through the mathematics to establish the roots of the closed loop transfer function from which closed loop frequency response plots can be obtained is both tedious and time consuming.

Simple graphical alternatives are available to the control systems engineer that are easy to use and provide a valuable insight into the relationship between the open loop and closed loop characteristics. The following paragraphs contain some significant mathematics that are included here for completeness in developing the concept of translating open loop plots to closed loop plots. It is not necessary for the reader to retain the mathematics provided here except to understand that there is mathematical logic behind the resulting graphics that are to be used in the analysis work that follows.

Returning to the Nyquist diagram we can say that the open loop plot may be represented by some gain term K in series with a number of dynamic terms represented by the term $G(j\omega)$ based on our approach to frequency response analysis where D is replaced by $(j\omega)$ in the transfer function. Therefore we can say that the open loop response of this system can be represented by the expression

$$\frac{\text{output}}{\text{error}} = KG(j\omega),$$

and the closed loop response by the equivalent expression

$$\frac{\text{output}}{\text{input}} = \frac{KG(j\omega)}{1 + KG(j\omega)}.$$

From our previous work we can represent the open loop response term $KG(j\omega)$ by a complex number of the form $a + jb$. From this it follows that the closed loop response term

$$\frac{KG(j\omega)}{1+KG(j\omega)} = \frac{a+jb}{[1+(a+jb)]}.$$

Since the expression on the right is simply a complex number we can say that lines of constant magnitude (gain) are simply the locus of the modulus of this complex number. Thus constant values of M can be determined from the equation:

$$M = \frac{|a+jb|}{|1+(a+jb)|}.$$

Rationalizing the above equation shows that the expression for lines of constant gain M are of the form:

$$\left[a - \left(\frac{M^2}{1+M^2}\right)\right]^2 + b^2 = \frac{M^2}{(1+M^2)^2}.$$

Those of you who remember your high school geometry may recognize the above expression as the equation for a circle which is of the form $x^2 + y^2 = r^2$. Similarly it can be shown that lines of constant closed loop phase angle N drawn on the Nyquist diagram can be represented by the expression

$$\left(a + \frac{1}{2}\right) + \left[b - \frac{1}{(2N)^2}\right] = \left(\frac{1}{4} + \frac{1}{4N^2}\right).$$

This equation also defines the loci of circles with the center at the coordinates

$$-\left(\frac{1}{2}, \frac{1}{2N}\right).$$

and a radius of

$$\pm\sqrt{\frac{(N^2+1)}{2N}}.$$

Figure 2.33 shows a Nyquist diagram with families of M and N circles superimposed. From these circles it is possible to extract the closed loop frequency response characteristics of an open loop Nyquist plot directly.

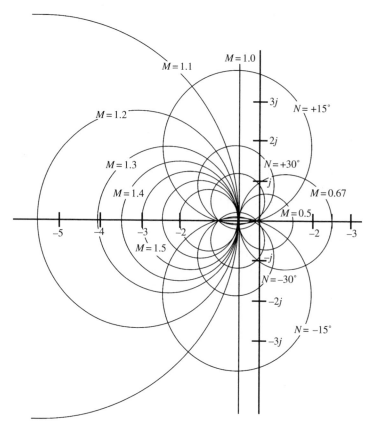

Figure 2.33 M and N circles

2.8.3 The Nichols Chart

There is, however, a fundamental problem with the Nyquist diagram that you may have already noticed. This is the fact that the use of amplitude ratio instead of its logarithmic equivalent gain in dB presents response loci where the frequency range of interest is 'squashed' tightly into the region between the instability point and the origin. Also the polar coordinate method of phase angle representation is not particularly user friendly.

The solution to this problem was developed by Nathaniel B. Nichols (1914–1997) who developed his own version of the Nyquist diagram with embedded M and N circles in a Cartesian coordinate form using gain in dB. This was first published in 1947 by James, Nichols and Phillips in a paper entitled '*Theory of servomechanisms*'. The extensive use of the Nichols chart as an important graphical tool today is a testament to the contribution of this paper to the field of feedback control theory. With the Cartesian coordinate format and the logarithmic scale of the gain axis (dB's) the M and N circles become a series of curves which when superimposed on the open loop gain versus phase graph become the Nichols chart shown in Figure 2.34.

Open loop phase moves to the left on the x-axis with increasing phase lag. Open loop gain is plotted as the y-axis value with zero dB being the horizontal axis in the center of the chart. The closed loop gain and phase can be obtained directly from the curved lines. Standard charts are available to use for plotting response curves. The figure here shows only a select few of the closed loop gain and phase curves in the interest of clarity. It may be helpful to consider the closed loop gain values as a third dimension coming out of the paper forming a 'mountain' of infinite height at the instability point.

In order to become familiar with the use of the Nichols chart, we will develop a number of examples. To begin let us consider the example of a single integrator $1/D$ in the forward path with a unity feedback path and no other dynamic terms as indicated in the block diagram of Figure 2.35. Note that the open loop transfer function (OLTF) for this example is $1/D$ and the closed loop transfer function (CLTF) is:

$$\frac{1}{(1+D)}.$$

This is a first-order lag with a time constant equal to 1.0 second.

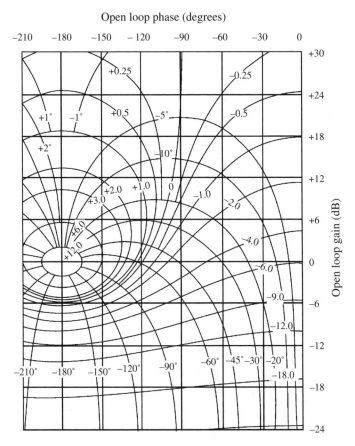

Figure 2.34 The Nichols chart

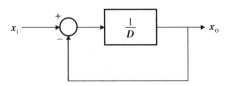

Figure 2.35 Single integrator example

The Nichols chart of Figure 2.36 shows the frequency response plot of this simple system. The open loop phase is constant at −90 degrees and since the integrator gain is 1.0, its response curve must cross the 0 dB line at a frequency of 1.0 radian per second. Observe now where the closed loop values cross this open loop plot. At the $\omega = 1.0$ radians per second point where the open loop gain is 0 dB, the closed loop values are −3 dB gain and −45 degrees phase lag as predicted from our previous first-order lag analysis. Doubling the frequency to $\omega = 2.0$ radians per second shows the open loop gain to be half the magnitude of the $\omega = 1.0$ radian per second value at −6.0 dB which corresponds to

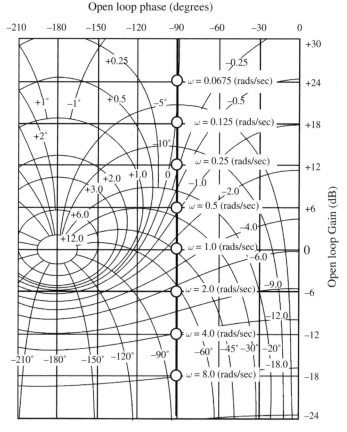

Figure 2.36   Nichols plot of OLTF 1/D

the closed loop values of −7 dB and −63 degrees of phase. This is again in agreement with our previous analysis.

Let us now go to a more complex example defined by our aircraft control system response generated in Section 2.7 and used in the Nyquist diagrams of Figure 2.31 and 2.32 earlier in this chapter. This same system response is shown on the Nichols chart of Figure 2.37. The most important observation to make here is the clarity of the information presented compared with the equivalent plot of the Nyquist diagram. Because of the logarithmic base, the whole frequency range of interest is visible while showing clearly the important region

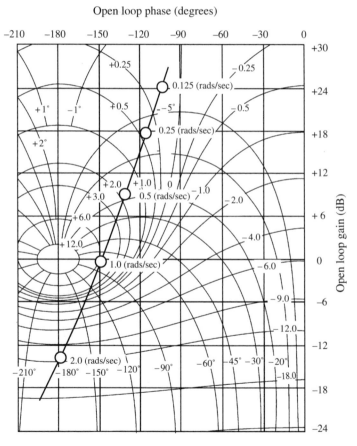

Figure 2.37 Nichols chart of the aircraft control system example

around the instability point. Again, the ability to derive the closed loop response immediately from the same plot is straightforward and insightful.

2.8.4 Graphical Methods – Summary Comments and Suggestions

This section has focused, so far, on the development of the Nyquist diagram and the Nichols chart as graphical methods for presenting control system frequency response data and being able to easily see the relationship between open loop and closed loop performance. The open loop response takes all of the elements around the control loop in series in order to determine the degree of stability of the system, in terms of gain and phase margins.

The Nyquist diagram was presented here as a stepping stone to the Nichols chart which embeds both open and closed loop lines using the gain (dB) convention. This graphical method is, in the opinion of the author, the tool of choice for closed loop control system analysis and synthesis. There can be some difficulty for the new user of this tool to overcome the visual complexity associated with all of the closed loop gain and phase curves superimposed on the open loop coordinates. This can sometimes make it difficult to observe what is really going on.

One technique, mentioned briefly earlier, that may be helpful to the control systems engineer is to focus on the closed loop gain curves of the Nichols chart and to visualize these gain values as a third dimension coming out the chart as interpreted by the artistic impression presented as Figure 2.38. As shown the closed loop gain curves become progressively higher as they approach the instability point This 'mountain' rising out of the page implies that a response curve passing close to the instability point (0 dB, −180 degrees open loop), as is the case when stability margins are small, will have to cross the 'mountain' at a 'high altitude' resulting in the closed loop gain response being greatly magnified for the frequencies corresponding to the high points on the 'mountain'. This effect is equivalent to the resonance displayed by the spring–mass system analyzed in the first chapter. As the response curve penetrates into the 'mountain' the system will become progressively more oscillatory when operating as a closed loop system until instability occurs as the instability point is crossed.

A second observation is worth noting here with regard to the Nichols chart. The lower half of the chart shows the open and closed loop gains

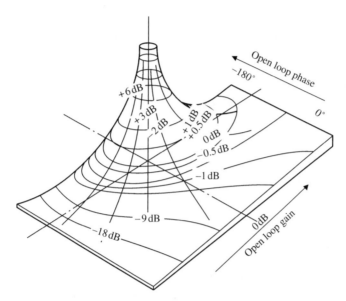

Figure 2.38 Three-dimensional interpretation of the Nichols closed loop gain curves

and phase angles becoming progressively closer together as the response becomes more and more attenuated. This makes sense because as the frequency of the command signal increases the system response becomes more delayed and attenuated to the point where the system output is very small. The effect of the feedback loop therefore becomes so small as to be negligible and thus the open and closed loop responses begin to coincide.

Earlier in this chapter we used separate plots of gain and phase angle against frequency to define the system performance. The open loop plot of this type is the traditional 'Bode diagram' where gain and phase margins can be observed. This same form of frequency response presentation is also used to show closed loop performance; however, in this case the graphs are simply 'frequency response plots' and while stability margins cannot be determined from these plots the degree of stability can be inferred simply by observing the peaking tendency of the gain curve. A system having a large magnification at a specific frequency (like the resonance of the spring–mass system) implies that stability margins are small.

There is a short cut approach to the determination of whether or not a closed loop system will have good stability margins using only the open loop gain versus frequency plot. This short cut rule states:

If the open loop gain plot crosses the 0 dB line with a slope of -6 dB/octave (or less) and this slope is maintained for about half a decade either side of the crossover point, the system will have good stability margins.

While the primary weakness of the Bode diagram and frequency response methods described here is the inability to easily translate between open and closed loop, this graphical method is by far the most popular method used to display open and closed loop system behavior and therefore the reader is encouraged to become familiar with its use.

A word about the term 'bandwidth' is also worth discussing here. This term refers to the transition frequency where a control system begins to become more and more attenuated and delayed relative to the input command. For a first-order lag this transition frequency is $1/T$ radians per second and for a second-order system the transition occurs around the undamped natural frequency ω_n radians per second. In each case this frequency indicates the region where the response curve plotted on the Nichols chart comes closest to the instability point. One decade or so either side of this frequency (or bandwidth) is known as the 'frequency range of interest' for that specific system. Frequencies lower than this range approximate to the steady state behavior of that system and frequencies above the range typically produce a highly attenuated response.

2.9 Chapter Summary

This chapter is the foundation of this book in that it develops the basic definition of the closed loop system and how the dynamic performance of closed loop control systems are analyzed and tested from the perspective of response and stability. The requirements for stable behavior of closed loop systems were established together with the definition of stability margins. This allows the control system designer to determine the quality of stability that can be expected and today serves as an effective way to specify the dynamic performance requirements of a system.

The integration process was singled out as one of the most important functions in closed loop control due to its ability to eliminate steady state errors and many examples were developed to explain its dynamic behavior. In particular, the quarter of a cycle of phase lag (in a response to sinusoidal input) that is produced by the integration process can become critical to the establishment of stable behavior of closed loop systems containing one or more integrating elements in the loop.

The decibel convention was introduced as a convenient way to linearize the gain versus log-frequency plots and many examples of using the frequency response approach to systems analysis were described. The most commonly used tool for expressing the dynamic characteristics of closed loop control systems, namely the frequency response plot, was described and rules for generating the gain and phase angle of typical elements and systems were developed and reinforced with examples. In its open loop form the frequency response plot is termed the 'Bode' diagram and illustrates the stability margins of the system. Frequency response plots are also a common medium for expressing the closed loop performance characteristics of closed loop systems.

The Nichols chart was introduced as a powerful graphical tool that can also be used to present frequency response characteristics. Here the gain and phase curves of a conventional frequency response plot are combined into a single locus with specific frequencies identified as points on the locus This tool, however, is unique in the fact that it allows the control engineer to see both the open loop response (with a clear definition of the stability margins) and the open loop characteristics from the same plot. The frequency response plot, together with the Nichols chart, are defined as the most important graphical aids in the analysis of closed loop systems and the chapters that follow only serve to reinforce this opinion.

3

Control System Compensation Techniques

3.1 Control System Requirements

The need for control system performance compensation is often dictated by the specified requirements for that system. Steady state accuracy requirements may dictate the need for the use of one or sometimes two integrators into the controller since this may be the only way to meet the specification. The resulting phase lag of 90 degrees for each integrator will often result in poor closed loop stability. In fact, the *class 2* system (with two integrators in the loop) is fundamentally unstable without some sort of dynamic compensation to ensure that the open loop response locus can circumvent the instability point at $|-1.0| \angle 180°$.

Stability requirements may be stated in the form of a gain and/or phase margin or as a maximum output overshoot following a set point step change. The control system may also be required to meet specific frequency response performance specified in terms of closed loop phase lag and/or gain boundaries at one or more frequencies.

The rationale behind these requirements is often related to the fact that one control system may be just one element in a larger and more

Stability and Control of Aircraft Systems: Introduction to Classical Feedback Control   R. Langton
© 2006 John Wiley & Sons, Ltd

encompassing system. In some cases, one control system requirement can be in conflict with another. For example, with regard to hydraulic actuation systems, stability and stiffness requirements often compete with each other. Stiffness, and its dynamic counterpart impedance, is one of the most challenging requirements to comply with. This is because nature always errs on the 'soft' side. Hydraulic oil absorbs air which lowers the effective bulk modulus (volumetric elasticity) to well below the advertised values shown in textbooks and material specifications. Also, bearings at attachment points have 'soft centers' further lowering the critical frequencies measured by test.

Ironically, the stability of actuation systems with high inertia loading can be eased by 'softening' the system so that it deflects more as a result of external loading. This can be achieved using some sort of pressure feedback. This type of compensation is a convenient way to provide pseudo acceleration feedback; however, in its simplest form, pressure (or force) feedback can result in a loss of control effectiveness when operating against high aerodynamic forces.

From this brief overview and example related to control system specifications, it can be appreciated that control system compensation can become difficult to optimize since an ideal solution to all of the functional requirements may not be possible and therefore some compromise may be necessary.

3.2 Compensation Methods

Control system compensation is the strategy used by the control system designer to improve system dynamic performance through the addition of dynamic elements in order to mitigate some of the undesirable features of the control elements present in the system. Such undesirable features may include:

- the integrator lag of 90 degrees,
- slow response of some transducers and sensors,
- process delays, non-linearities and other undesirable characteristics.

Control system compensation invariably involves the introduction of 'anticipation' into the control loop. This is accomplished using differentiation (measuring the rate-of-change) of control signals around the control loop. For example, in a position control loop the position feedback signal could be differentiated to obtain the velocity of the output. This velocity signal can then be used to anticipate the position of the output at some future time. Anticipatory devices can be 'built' by combining transfer functions to achieve the desired control action. Several examples are presented in the following paragraphs.

3.2.1 Proportional Plus Integral Control

From the previous work we have noted the benefit of integral action in a closed loop system as a means of ensuring that the steady state error is reduced to zero as a result of its inherent high gain at low frequency (infinite at zero frequency). The 90 degree phase lag that accompanies this device, however, can be troublesome in the transition frequency range where the open loop gain crosses the 0 dB line and stability margins are established. One way to circumvent this problem is to introduce a proportional path into the error signal line such that the proportional and integral components are developed in parallel and their individual outputs summed as shown in Figure 3.1.

The output from this proportional plus integral control arrangement is simply the error multiplied by $\left(\frac{K_I}{D} + K_P\right)$ which is the sum of the integral and proportional path contributions. Intuitively it can be seen that at high frequencies, the integrator will become greatly attenuated (because the D term, i.e. $j\omega$, becomes large) and therefore the proportional path will dominate the control output. Similarly at very low frequencies, the integrator will have a high gain and therefore its output will be the dominant term in the control output.

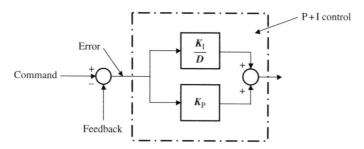

Figure 3.1 Proportional plus integral control

The above expression can be rationalized as follows:

$$\left(\frac{K_I}{D} + K_P\right) = \frac{K_I\left(1 + \frac{K_P}{K_I}D\right)}{D} \text{ or } \frac{K_I(1 + TD)}{D}$$

where T is a time constant K_P/K_I.

We have now created a first-order term in the numerator which is called a 'first-order lead' (as opposed to a lag).The frequency response of a first-order lead term is the inverse of the lag in that the phase angle and gain increase with increasing frequency.

Another way to develop this form of transfer function is shown in Figure 3.2. Note that the feedback in the figure is positive, therefore to rationalize this system into a single transfer function we must use the rule: forward path divided by one *minus* the loop. Referring to the figure this yields:

$$\frac{1}{1 - \dfrac{1}{(1 + TD)}} = \frac{(1 + TD)}{D}.$$

This is exactly the same expression that we developed by rationalizing the initial P + I control arrangement. The interesting aspect of this demonstration is that integral action can be generated without actually using an integrator and this technique is used frequently by the control engineering community. While it is usually simple with electronics to put together any desirable control function, it is often difficult in other disciplines such as hydraulics, pneumatics and mechanics. Here it is usually much easier to build a delay using restrictors and compliant volumes to emulate a first-order lag than to develop pure integral action.

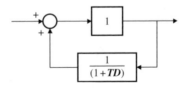

Figure 3.2 Alternative P + I implementation

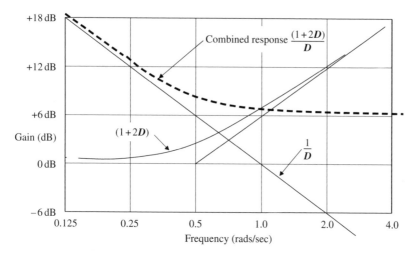

Figure 3.3 Proportional plus integral gain plot

In order to reinforce the control characteristics of the P + I control, let us develop the frequency response curves using specific values: $K_I = 1.0$ and $K_P = 2.0$. These values yield the following control transfer function:

$$\frac{(1+2D)}{D}$$

where the 2 second time constant in the numerator is K_P/K_I.

Figure 3.3 shows the gain of this function plotted against frequency and Figure 3.4 shows the phase angle. From the above plots it can be seen that at low frequencies the integration term dominates with the control action resembling a simple integrator with a gain of 1.0 s^{-1} (or radians per second). At frequencies above about 2.0 radians per second, however, the control action is almost entirely equal to a proportional gain of 2.0 (which is +6 dB) with essentially zero phase angle.

The P + I control action therefore seems to provide the best of both worlds, i.e. integral control at low frequency and steady state to ensure zero steady state error and proportional control at the higher frequencies without the burden of the −90 degree phase lag. Selection of the integral and proportional gains allows the control systems engineer to position the lead time constant to cancel out a major lag term in the control loop, e.g. in the process itself, that cannot otherwise be easily modified.

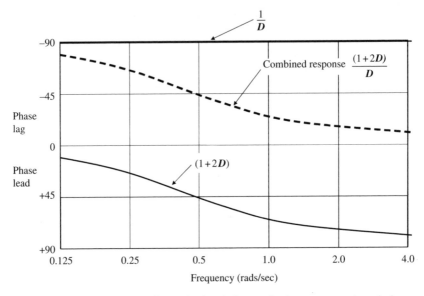

Figure 3.4 Proportional plus integral phases angle plot

3.2.2 Proportional Plus Integral Plus Derivative Control

An extension of the P + I example is the three-term controller having P + I + D control action, the third term representing derivative control action. Figure 3.5 illustrates this control concept schematically. This approach adds another level of capability (and complexity) to the control action. Rationalization of the controller transfer function as before yields the following:

$$\frac{K_I}{D} + K_P + K_D D = \frac{K_I}{D}\left(1 + \frac{K_P}{K_I}D + \frac{K_D}{K_I}D^2\right).$$

We now have a second-order term in the numerator which can be represented by either two first-order terms in series or a second-order term with imaginary roots depending upon the values chosen for K_I, K_P and K_D.

For real roots we have two first-order lead terms that can be designed to cancel major lag terms in the control loop as before in the P + I example. For the imaginary roots solution we can treat this just as we did the oscillatory spring–mass system defining values for the natural

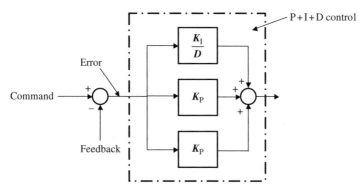

Figure 3.5 Proportional plus integral plus derivative control

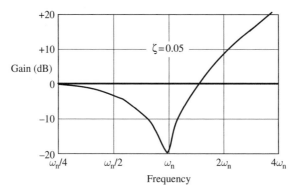

Figure 3.6 Second-order numerator gain response showing notch at ω_n

frequency ω_n and the damping ratio ζ. The fact that this term is in the numerator means that the frequency response is the inverse of that developed in Chapter 1 with both gain and phase increasing with frequency. A gain plot of a numerator second-order system with a damping ratio of $\zeta = 0.05$ is shown in Figure 3.6.

From this figure it can be seen that the inverted resonance effect at ω_n can be positioned along the frequency axis to cancel out a troublesome resonance in the denominator of the open loop transfer function associated with the process or other control loop element. If the element to be neutralized has a very low damping ratio, this type of compensation (which is referred to as 'notch filtration') requires great precision in positioning the natural frequency of the compensation device relative to the natural frequency of the denominator term to be cancelled. Even small errors in the design of the notch filter can result in a substantial loss of effect in terms of overall system performance and stability.

For most control system applications the use of three-term controllers may be regarded as an 'overkill' because the problems introduced by the derivative term can more than offset its benefits because the derivative term can be a major source of noise generation within the control system. The differentiation process is inherently noisy and can substantially magnify even small levels of noise to a level that can seriously affect the performance of the system particularly when noise frequencies are high. It is therefore good practice when using a three-term controller with derivative action to include a high frequency filter to protect the system against unwanted high frequency noise. The application of three-term controllers is more common in relatively slow response systems such as is typical of the process control industry.

3.2.3 Lead–Lag Compensation

Lead–lag compensation is in effect lead compensation with a high frequency lag term to filter out high frequency noise. In other words it is the lead term that we want to introduce into the system in order to cancel out lag elements in the control loop over the critical range of frequencies where stability margins are established.

We must be careful, however, not to introduce unwanted high frequency noise into the system due to the differentiator in the numerator. Lead–lag compensation is the resulting compromise. The transfer function of this form of compensation is of the form:

$$\frac{(1+T_1 D)}{(1+T_2 D)}.$$

Here the time constant in the numerator is selected to compensate a lag in the control loop while the denominator break frequency $1/T_2$ should be well beyond the frequency range of interest. Typically the

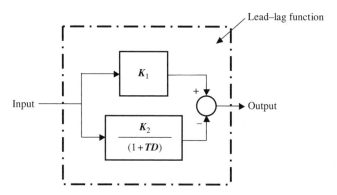

Figure 3.7 Lead–lag schematic

denominator time constant is at least an order of magnitude smaller than the time constant in the numerator. We can 'construct' the lead–lag function using the arrangement shown in Figure 3.7.

From this figure we can develop the transfer function for the device as follows:

$$\frac{\text{output}}{\text{input}} = \frac{K_1\,(1+TD) - K_2}{(1+TD)} = \frac{(K_1 - K_2)\left[1+\left(\dfrac{K_1}{K_1 - K_2}\right)TD\right]}{(1+TD)}.$$

If we select $K_1 = 10$, $K_2 = 9$ and $T = 1.0$ into the above transfer function we obtain:

$$\frac{\text{output}}{\text{input}} = \frac{(1+10D)}{(1+D)}.$$

We now have a numerator lead term with a time constant of 10.0 seconds and a denominator lag of 1.0 second. The ratio between the numerator and denominator time constants can be adjusted by simply modifying the values of K_1 and K_2.

The frequency response plots for the gain and phase angle of this function are shown in Figure 3.8. From the frequency response plots we can make the following observations.

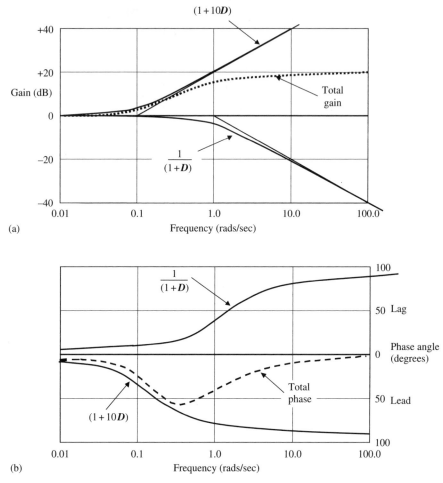

Figure 3.8 Lead-lag function (a) gain and (b) phase response

- The lead–lag function has unity transmission in steady state, i.e. 0 dB and 0 degrees phase shift.
- Between the two break frequencies the phase angle goes positive, i.e. there is a 'phase lead' or 'phase advance' response. This phase lead reaches a maximum value, which is a function of the separation between the two break frequencies, before reducing back towards a zero phase angle as the frequency is increased above the second break frequency.

- The separation between the two break frequencies must be at least a factor of 10 to be effective, i.e. lower separation multiples have relatively small phase lead excursions and therefore little effect on the overall system behavior.

The lead–lag compensation strategy is often used to offset the effect of slow response transducers as demonstrated in the example in Section 3.3.

3.2.4 Lag–Lead Compensation

Lag–lead compensation is really a poor man's P + I control since it produces essentially the same dynamic effect as the P + I compensation described in Section 3.2.1. Lag–lead is accomplished by introducing a first-order lag with a very long time constant together with a high gain. In the frequency range of interest this looks very similar to an integrator with phase lag approaching 90° together with a gain slope of $-6\,\mathrm{dB}$ per octave. The lead term cancels the lag effect as the critical frequency range is neared as is the case with the pure proportional plus integral control action. The lag–lead compensation transfer function, therefore, is of the form:

$$\frac{K\,(1+T_1D)}{(1+T_2D)}\ .$$

As described above, K is typically very large and T_2 is also large so that it looks like integral action in the frequency range of interest. T_1 is chosen to cancel a major system lag term as in the P + I compensation approach described previously. The construction of the lag–lead network is similar to the lead–lag method (see Figure 3.9). The values for K_1, K_2 and T are substantially different from the lead–lag example and the two paths are summed to obtain the output rather than differenced.

The resultant transfer function is:

$$\frac{(K_1+K_2)\left[1+\left(\dfrac{K_1}{K_1+K_2}\right)TD\right]}{(1+TD)}\ .$$

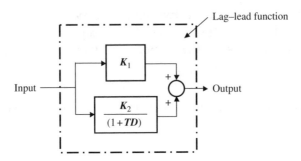

Figure 3.9 Lag-lead schematic

To demonstrate this lag–lead function let us select $K_1 = 1.0, K_2 = 99.0$ and $T = 10.0$. Substituting these values into the above transfer function yields the lag–lead control action:

$$\frac{\text{output}}{\text{input}} = \frac{100.0\,(1+0.1D)}{(1+10D)}.$$

Once again this transfer function can be 'built' using the positive feedback of a first-order lag (similar to the example given in the proportional plus integral compensation in Section 3.2.1) as indicated in Figure 3.10. Rationalizing the block diagram of Figure 3.10 into a single transfer function by applying the rule 'forward path divided by (1-the loop)' we obtain:

$$\frac{1}{1-\left[\dfrac{0.99}{(1+0.1D)}\right]} = \frac{(1+0.1D)}{(1+0.1D)-0.99)}$$

$$= \frac{100\,(1+0.1D)}{(1+10D)}.$$

This is the same transfer function previously developed.

Figure 3.11 shows the gain and phase frequency responses for the above lag–lead function showing the clear similarity between the lag–lead compensation and the pure proportional plus integral control action.

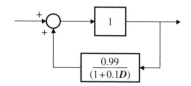

Figure 3.10 Alternative lag–lead approach

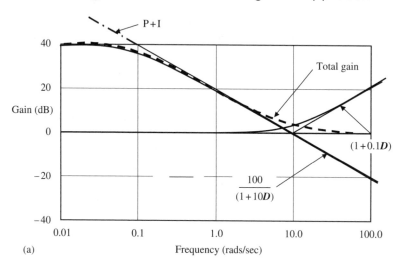

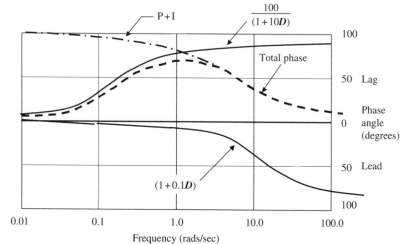

Figure 3.11 Lag–lead (a) gain and (b) phase response

We can make the following observations regarding lag–lead compensation.

- Lag–lead provides a pseudo P+I type of control action
- In steady state the gain of the lag–lead compensator is very large but finite and this is the main functional difference between lag–lead and P+I and will result in a small but measurable steady state error.
- In the example shown in Figure 3.11(a) and (b) the response is almost identical to the pure P+I compensation of $\left(1+\frac{10}{D}\right)$ for frequencies above the lag break frequency.

3.2.5 Feedback Compensation

This approach to improving the dynamic behavior of feedback control systems can be extremely useful if the forward-path transfer function is highly non-linear, varies in a unpredictable manner, or has other undesirable characteristics that make traditional forward path compensation ineffective. The success of this compensation approach is due to the simple fact that we can introduce feedback transfer function elements that are fixed with known and reasonable values.

The reasoning behind the concept of feedback compensation is explained by considering the simple block diagram of Figure 3.12 which shows a typical closed loop system with forward path and feedback elements.

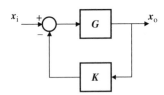

Figure 3.12 Simple closed loop system

We can express the closed loop transfer function (CLTF) of this system as:

$$\frac{x_o}{x_i} = \frac{G}{1+KG}$$

which is the same as:

$$\frac{1}{K}\left(\frac{KG}{1+KG}\right) \text{ or } G\left(\frac{1}{1+KG}\right).$$

Now if $KG \gg 1$ for the particular signal range being considered then, based on the first of the above two expressions we can say:

$$\frac{x_o}{x_i} \simeq \frac{1}{K}.$$

This assumption will usually apply to the low frequency range of input signals.

For the conditions where $KG \ll 1$ which would typically correspond to high frequencies when significant attenuation is present, the closed loop transfer function approximates the forward path transfer function G based on the second of the above expressions. Hence using this technique the low frequency response can be approximated by the inverse of the feedback transfer function. This will transition to the forward path response at higher frequencies as KG attenuates.

To illustrate this compensation technique let us consider the closed loop control system shown in Figure 3.13 which describes a system with a process which has an undesirable transfer function in the form of a very lightly damped second-order system. The second-order term introduces an almost instantaneous phase lag of 180 degrees at the resonant frequency. This severely limits the integrator gain K_1 that can be

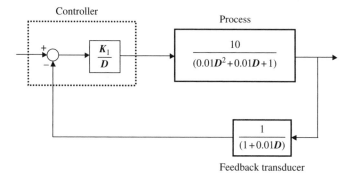

Figure 3.13 Control system with undesirable process dynamics

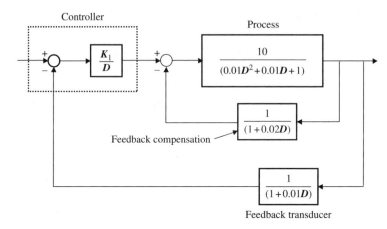

Figure 3.14 Control system with feedback compensation

used resulting in a system that is either too sluggish in response or too oscillatory due to the process characteristics. This problem may be compounded substantially if the predictability of the process resonant frequency is not very reliable or varies over the operating range of the control system.

Using the feedback compensation technique, this system is modified as shown in Figure 3.14 to include an additional feedback loop containing a first-order lead term with a time constant of 0.02 seconds. Rationalizing just the inner loop, the closed loop transfer function is:

$$\text{CLTF} = \frac{\dfrac{10}{(0.01D^2 + 0.01D + 1)}}{1 + \dfrac{10\,(1 + 0.02D)}{(0.01D^2 + 0.01D + 1)}} = \frac{1}{0.1\,(0.01D^2 + 0.01D + 1) + 1 + 0.02D}.$$

This further rationalizes into a second-order system with significantly improved damping as shown in the final equation below which shows the second-order system in the standard form indicating the undamped natural frequency and the damping ratio:

$$\left(\frac{1}{1.1}\right)\left\{\frac{1}{\left[D^2 / (33.16)^2\right] + \left[(D)\,(0.45)\,/33.16\right] + 1}\right\}.$$

The effect of the added feedback loop has been to increase the undamped natural frequency of the process from 10 to 33.16 radians per second and, at the same time, improving the damping ratio from 0.05 to 0.45. It should also be noted that the process gain has been reduced by a factor of about 10. This can be compensated for by increasing K_1 to achieve optimum stability margins.

According to the feedback compensation methodology, the inner loop response for the condition where the product of the forward path and feedback elements is significantly greater than unity can be approximated by the reciprocal of the feedback element. For the opposite condition where the product of the forward path and feedback elements is significantly less than unity, the process approximates to the forward path transfer function.

Figure 3.15 demonstrates these assumptions showing that the low frequency gain response does indeed look like a first-order lag with a time constant of 0.02 seconds which is precisely the reciprocal of the feedback element. This transitions into the forward path response at the higher frequencies where the product of the forward path and feedback elements becomes significantly less than unity.

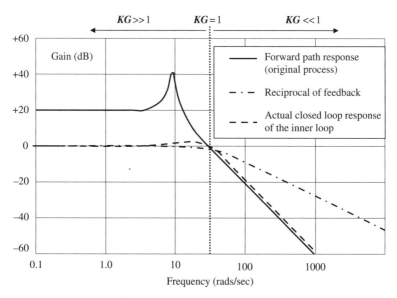

Figure 3.15 Improved inner loop performance

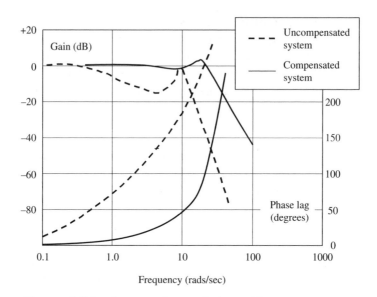

Figure 3.16 Comparison of closed loop responses

In this simple example we can do the mathematics to confirm the conclusions arrived at above because of the simple linear example used to demonstrate the feedback control principle. The intent here is to prove the concept by example so that when presented with complex non-linear processes, the control engineer can apply the feedback control methodology with both confidence and understanding. In the above example the complete system can be optimized by the selection of the controller gain K_1 to provide a much improved bandwidth together with good stability margins. Figure 3.16 compares the original uncompensated system with the best controller gain with the same system with the feedback compensation added. The uncompensated system can only tolerate an integral gain of 0.1 which yields only 6.0 dB of gain margin and a sluggish response. In contrast, the compensated solution has an integral gain of 10.0, a much faster response and even better stability margins.

So far we have covered five commonly used compensation techniques for improving the dynamic performance characteristics of closed loop control systems. There are many others not covered here and the reader is encouraged to experiment with the building block approach to build compensation transfer functions. Furthermore, compensation using second-order transfer functions, which was covered here only briefly, can be particularly valuable when presented with a moderately resonant but predictable element or process that impacts the stability margins as a result of the added phase shift in the critical frequency range. This situation will be addressed in the 'Class 2' design example described later.

3.3 Applications of Control Compensation

This section applies some of the compensation concepts described above to specific control system examples in order to demonstrate the performance benefits that can be accomplished via this technique.

3.3.1 Proportional Plus Integral Example

Let us examine a real world control system application to demonstrate the benefits of P + I control over pure integral control action. The example that we will use in based on an auxiliary power unit (APU) for an aircraft that is used to provide electrical power to the aircraft when the engines are shut down during turn-around at the gate. The APU also provides a source of compressed air for engine starting. Figure 3.17 shows the APU schematic focusing on the fuel control arrangement that is used to control the shaft speed of the APU.

The control challenge here is to ensure that when running, the APU shaft speed remains essentially constant so that the electrical power generator maintains constant frequency ac power supply to the aircraft (usually 400 Hz). To achieve this, the controller must contain an integrator to ensure that the speed error will be zero under all APU power generator loads.

Note that if we use a simple gain term for the controller, some small error is necessary to generate a finite fuel flow to run the APU. As the power demanded from the APU increases, additional fuel is required and with a simple gain controller, this can only be provided at the expense of a larger error signal. This simple gain type of speed regulator is called a 'droop governor' which is descriptive of the response of this type of governor to increasing load. When precise speed regulation with essentially zero speed change over the complete load range is required, integral action becomes necessary and this type of speed regulator is referred to as an 'isochronous governor'.

Figure 3.18 shows the control system block diagram for the speed control loop showing a pure integral controller and the fuel metering valve represented by a simple gain term. The engine dynamics comprises a gain term together with two first-order lags. One lag is

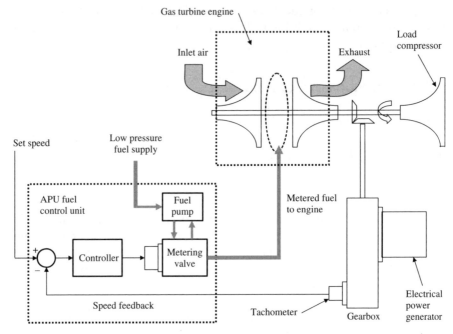

Figure 3.17 APU schematic

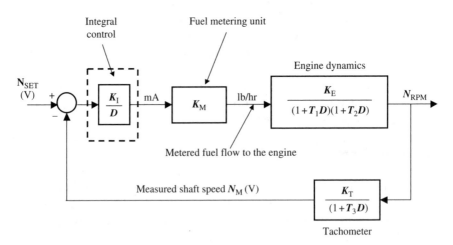

Figure 3.18 APU speed control loop block diagram

associated with the combustion process and the second is related to the inertia of the rotating hardware as torque is applied and speed response lags behind. The speed sensor is shown as a gain term and a first-order lag.

The control loop shows that the required operating shaft speed N_{SET} is compared with the tachometer speed transducer output to obtain the speed error which is then acted on by the integrator control shown in the dotted line box. The fuel metering valve is an electro-hydraulic servo valve that is assumed to have negligible dynamics for the purpose of this example. This valve converts the integrator output to fuel flow via a simple gain term.

Let us now allocate values for the variables and their units:

K_I is to be determined by our analysis and will have units of milliamps/second per volt of speed error (mA/sec V) thus as long as a speed error exists the integrator output will continue to increase (or decrease) until the error is zero

K_M converts milliamps to fuel flow and has units of pounds/hour per milliamp and for this example has a value of 4.0 lb/hr mA.

K_E is the engine gain with a value of 70 RPM/(lb/hr) at the operating condition under study.

K_T converts engine shaft RPM to an equivalent voltage and is allocated a value of 0.00025 V/RPM. (For an engine operating shaft speed of 40 000 RPM the tachometer output will be 10 V)

The time constants for this analysis are as follows: $T_1 = 0.02$ seconds, $T_2 = 1.0$ seconds and $T_3 = 0.1$ secconds.

This indicates that the dominant lag in the control loop is the engine inertia lag of 1.0 second. The other two lags are smaller but not small enough to neglect their contribution to the control loop dynamics. Let us now check the loop gain and ensure that it has the correct units.

$$\text{loop gain} = K_I (K_M K_E K_T) = K_I (4.0) (70.0) (0.00025) = 0.07 K_I$$

Units around the loop are:

$$\frac{\text{mA}}{\text{sec V}} \cdot \frac{\text{lb/hr}}{\text{mA}} \frac{\text{RPM}}{\text{lb/hr}} \frac{\text{V}}{\text{RPM}} = \text{sec}^{-1}.$$

As you can see, when everything is cancelled we are left with $\sec^{-1}$ or radians per second which is correct for a control loop with one integrator.

Based on the above variable values we can define the open loop transfer function (OLTF) as:

$$\frac{N_M}{(N_{SET} - N_M)} = \frac{0.07K_I}{D}\left[\frac{1}{(1+0.02D)(1+D)(1+0.1D)}\right].$$

We can now figure out what value of K_I will yield acceptable stability margins. Using the short cut method of plotting only the gain curve against frequency we can estimate what value of K_I will keep the rate of attenuation at 6 dB/octave for about a decade in the region where the gain curve crosses the 0 dB line. Figure 3.19 shows the open loop gain response for the above function excluding the integrator term. By inspection, if we add an integrator to the gain plot of Figure 3.19 and require that the total gain maintains an attenuation rate of 6 dB per octave for about a decade around the crossover point, the maximum value of the integrator term $0.07\,K_I$ we can select is about 0.5 which defines the integrator crossover frequency in radians per second.

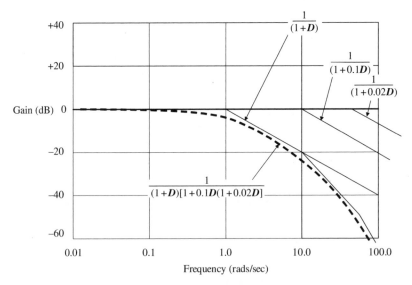

Figure 3.19 Open loop gain without the integrator term

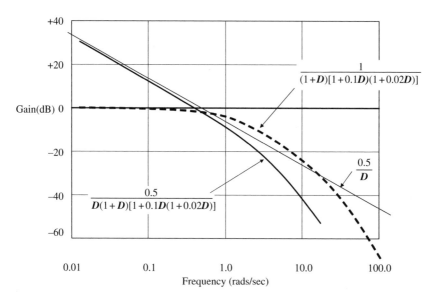

Figure 3.20   Open loop gain with the best integrator value

Figure 3.20 shows the total gain plot including the integrator. In this plot the integrator gain K_I is defined by the requirement that $0.7\,K_I = 0.5$.

Let us now examine how the same system can have substantially improved dynamic performance by using $P + I$ control in place of the pure integral action evaluated above. From the $P + I$ description presented in Section 3.2.1 we know that the transfer function for this type of control action can be represented as:

$$\frac{K_I\,(1 + TD)}{D}$$

where the time constant $T = K_P/K_I$.

The open loop transfer function for the speed control loop with $P + I$ control is as follows with the known variable values inserted:

$$\frac{N_M}{N_{SET} - N_M} = \frac{0.07K_I\,(1 + TD)}{D\,(1 + 0.02D)\,(1 + D)\,(1 + 0.1D)}.$$

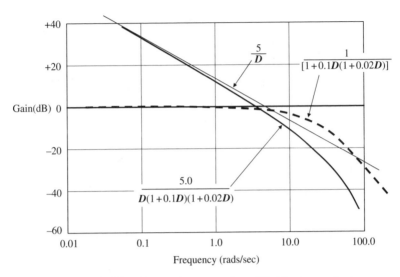

Figure 3.21 Open loop gain with P+I control

We can now choose the value of T to cancel one of the terms in the denominator. In this case setting $T = 1.0$ would exactly cancel the engine 1 second lag.

If we now go through the same exercise as for the integral controller we will see that we can select a value of 0.07 $K_I = 5.0$ and still provide good stability based on the gain slope around the crossover frequency. This is illustrated in Figure 3.21 which shows the composite open loop gain for the system.

Because we have been able to neutralize the primary lag in the APU transfer function, the control system with P + I control action is 10 times more responsive to dynamic disturbances than the system with pure integral control. The improved dynamic response of the compensated System is important because deviations in electrical power supply frequency variations following sudden changes in power demand are usually limited by industry specification requirements. Both design solutions are shown in the Nichols chart of Figure 3.22.

In this Nichols plot the response trajectory for both systems is essentially identical and is therefore represented here by a single curve. The difference between the systems, however, is in the frequency scale. The integral control system crosses the 0 dB line at 0.5 radians per second while the P + I system crosses the 0 dB line at 5.0 radians per second.

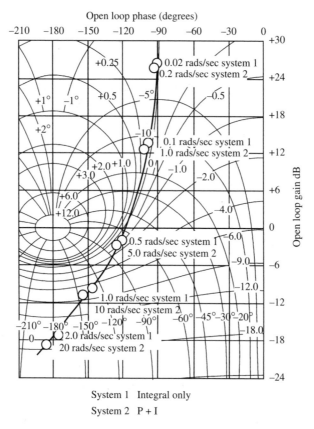

Open loop phase (degrees)

Figure 3.22 Nichols plot of both APU control system designs

This is easier to see in Figure 3.23 which shows the closed loop response curves for each system derived from the Nichols chart. Here it clearly shows that the P + I controlled system has a flat gain response out to about 5 radians per second while the integral controlled version begins to attenuate at about 0.5 radians per second.

An important point to note regarding the translation from open to closed loop via the Nichols chart is that the closed loop transfer function is assumed to have unity feedback with all of the control system elements in the forward path. In other words the chart translates from $KG\,(j\omega)$ to $KG\,(j\omega)\,/1 + KG\,(j\omega)$.

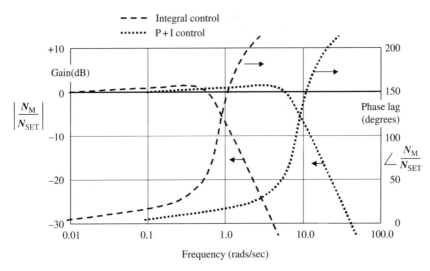

Figure 3.23 Closed loop frequency response for both solutions

In the example just completed the closed loop response is from the speed set point, N_{SET} to the output of the speed sensor N_M. Thus the speed tachometer is assumed to be in the forward path. In this case, therefore if we want the actual closed loop speed response of the system with the speed tachometer in the feedback path, i.e. $\frac{N}{N_{SET}}(j\omega)$ we must multiply the overall closed loop response by the inverse of the speed tachometer transfer function.

In mathematical terms we can say:

$$\frac{N}{N_{SET}}(j\omega) = \frac{N_M}{N_{SET}}(j\omega)\frac{N}{N_M}(j\omega).$$

While this looks complicated it is in fact quite simple. In this case we subtract the gain and phase of the first-order lag representing the speed tachometer to obtain the actual speed response as indicated in Figure 3.24. This figure compares the actual and the measured speed responses for the P+I control approach showing that the actual speed response is improved a little over the measured response as would be expected. It is important to note, however, that the stability of the closed

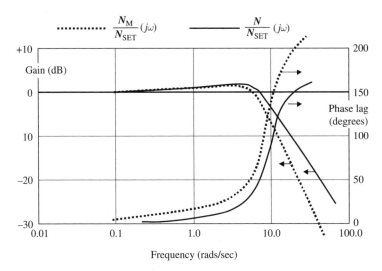

Figure 3.24 Actual versus measured speed response

loop system is exactly the same in both cases because the open loop response (and hence the characteristic equation) is the same.

The lessons learned from this example are clear. Pure integral control burdens the system with its 90 degree phase lag at all frequencies and thus limits the gain that can be selected while P + I control allows the control system designer to provide integral action at low frequencies while providing fast response at the higher frequencies without compromising stability margins.

An observation worth mentioning here refers to the 'short cut' approach to system design. In the example presented the short cut method gave conservative solutions with a gain margin of almost 20 dB and phase margin of over 50 degrees. While this method demonstrates a fairly conservative rule of thumb, designing for specific margins based on pre-determined system performance drivers will require a complete open loop response analysis. The short cut method, however, is an excellent place to start.

3.3.2 Lead–Lag Compensation Example

To demonstrate the application of lead–lag compensation we will use a fuel temperature control system example. This system, shown in the

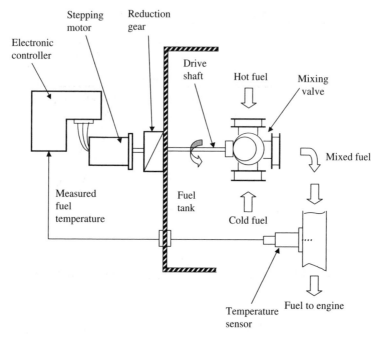

Figure 3.25   Fuel temperature control system schematic

schematic diagram of Figure 3.25 serves to control the temperature of the fuel that is fed to a gas turbine engine by mixing cold fuel from the storage tanks with hot fuel from heat exchanger discharge. Heat exchangers using fuel as a cooling medium are often used as a heat sink for aircraft systems that generate substantial amounts of heat such as hydraulics and avionics. By maintaining the engine feed fuel at a maximum allowable value, the availability of cold fuel as a heat sink is maximized.

In the example schematic an electronic controller compares the temperature set point with the measured engine fuel feed temperature. The error drives a stepping motor connected to a rotary ball valve having hot and cold fuel inlets. This valve serves as a mixing valve whose position determines the engine fuel feed temperature. The temperature of the discharge from the mixing valve is measured and fed back to the electronic controller.

Let us characterize each control loop element and construct a system block diagram. The temperature sensor has an inherent time lag equivalent to a first-order lag of 1.0 second and develops an output voltage 0.0 to 10.0 V for fuel temperatures from 0.0 to 100 C We can therefore represent this element with the transfer function $0.1/(1 + 1.0D)$.

The electronic controller compares the temperature sensor output with the temperature set point and drives the stepping motor at a stepping rate proportional to the magnitude of the error. A 10.0 degree error (i.e. 1.0 V) generates a stepping rate of 2500.0 steps per second. There is also a lag associated with the drive electronics and the stepping motor winding inductance of 0.05 second. We can represent the electronics and stepping motor by the transfer function $2500/D(1 + 0.05D)$ (motor steps per volt error). The motor rotates 10.0 degrees per step and the reduction gear ratio is 125:1 therefore there are 0.08 degrees of drive shaft rotation per degree of motor rotation. For this exercise we will ignore the effects of inertia and friction.

In order to continue with the analysis we need to characterize the mixing valve element at the specific operation condition to be analyzed. Figure 3.26 shows the results taken from a test rig. Here we can see that the slope of the line at an 80.0 C operating point is approximately 0.5 C of fuel temperature per degree of shaft rotation.

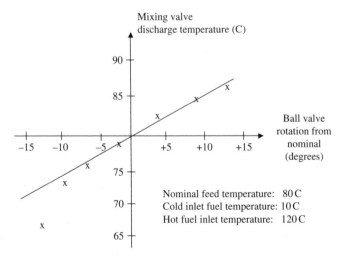

Figure 3.26 Mixing value sensitivity at 80 C

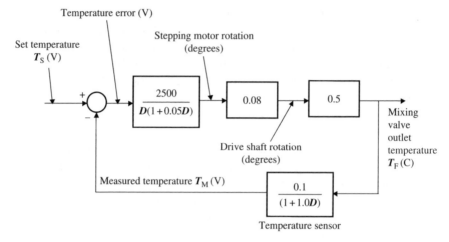

Figure 3.27 Temperature control system block diagram

From this we can now construct a system block diagram (see Figure 3.27) and determine the system stability margins. From the block diagram we can see that the open loop transfer function is:

$$\frac{(2500)\,(0.08)\,(0.5)\,(0.1)}{D\,[1+0.05D\,(1+1.0D)]} = \frac{10}{D\,[1+0.05D\,(1+1.0D)]}.$$

By generating the gain and phase for each element as before and adding them to arrive at the total gain and phase for the open loop transfer function we obtain the Bode diagram for the system. This is shown in Figure 3.28 and it is clearly apparent that the stability margins are unacceptably low. The phase margin in particular (the additional phase lag to achieve a loop phase lag of 180 degrees when the open loop gain passes through 0 dB) is only about 10 degrees or less.

The stability margins can be substantially improved by compensating for the 1 second first-order lag of the temperature sensor using a lead–lag function. We want the numerator of this function to cancel the sensor lag and if we choose the separation between the lead and lag time constants to be 50:1 we obtain the lead–lag transfer function $(1+1.0D)/(1+0.02D)$. The compensated system open loop transfer function then becomes $10/D\,[1+0.05\,D\,(1+0.02D)]$.

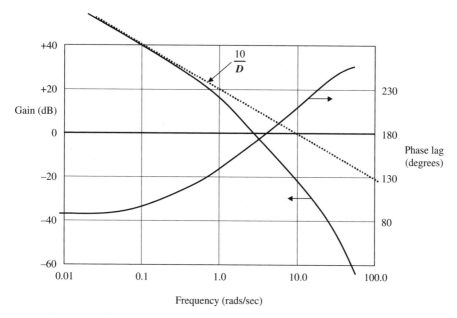

Figure 3.28 Temperature control system Bode diagram

Figure 3.29 is a Nichols chart showing both the uncompensated and compensated system responses. This chart shows clearly the substantial improvement in stability margins. The compensated system phase margin is increased to almost 60 degrees. Note also that the frequency at which the response curves cross the zero dB gain line is considerably higher for the compensated system implying that the closed loop bandwidth and hence the dynamic performance is improved.

To illustrate the improvement in dynamic response Figure 3.30 shows the closed loop frequency response for the two systems as derived from the Nichols chart. The uncompensated system has a sharp resonance at about 3 radians per second, while the compensated system shows flat response to about 10 radians per second implying a system substantially more capable of handling dynamic disturbances.

3.3.3 Class 2 System Design Example

Class 2 systems, i.e. systems having two integrators in the loop, are inherently unstable without some form of compensation to bring the open loop frequency response locus around the 1.0, −180 degrees

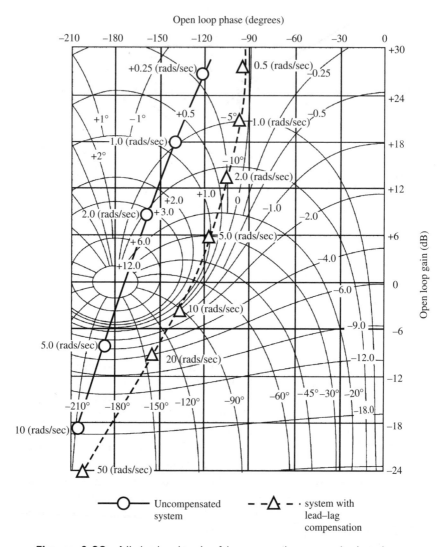

Figure 3.29   Nichols chart of temperature control system

point on the stable side as illustrated in the Nichols chart sketch of Figure 3.31.

The requirements that drive the need for a class 2 control system solution are for zero steady state error when following a command signal

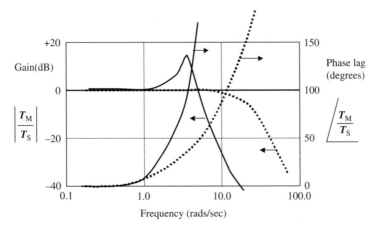

Figure 3.30 Closed loop response compensated and uncompensated

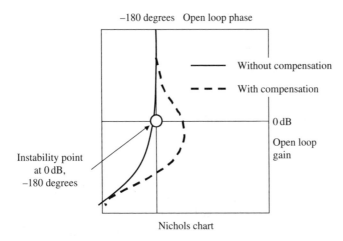

Figure 3.31 Typical class 2 open loop response

that is changing at a constant velocity. A class 1 system will ensure zero steady state *position* error while a class 2 system ensures zero *velocity* error. This is illustrated by Figure 3.32 which shows ramp responses for class 0, class 1 and class 2 systems.

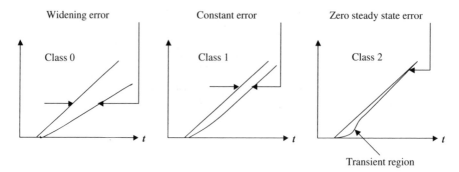

Figure 3.32 Ramp responses for class 0, 1 and 2 systems

The example developed here is for a target tracking system that requires the ability to follow a target moving at a constant velocity with zero error. The specified performance requirements are:

(1) zero error when following targets moving at constant velocities;
(2) maximum error when following a sine wave of $+/-25$ degrees at 0.5 radians per second shall be less than 1.0 minute of arc;
(3) small signal step change overshoot shall be less than 40 %.

Control System Description

Figure 3.33 shows the control system in schematic form. This control system uses an electronic controller to drive a hydrostatic drive comprising a variable displacement hydraulic pump and a fixed displacement hydraulic motor. The rotational output from the hydraulic motor drives the mechanical load and transmission. An electrohydraulic servo actuator controls the displacement of the pump in order to vary the speed and torque output of the motor. The transmission gearbox is coupled directly to the hydraulic drive motor.

Interpretation of Requirements

From the specification, we can make a number of design decisions that must be met in order to satisfy the peformance requirements. From

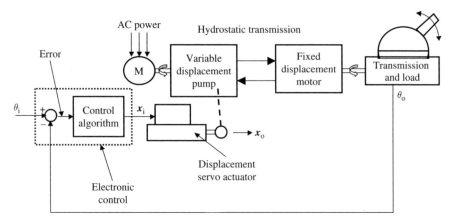

Figure 3.33 Tracking system schematic

requirement number (1) we must have two integrators between the error and the output as indicated by the equation:

$$\text{output}\,(\theta_{\text{o}}) = \frac{K_{\text{L}}}{D^2}\ \text{error.}$$

where K_{L} is the open loop gain. We can determine the minimum value of the gain K_{L} required to meet the specified performance from the equation:

$$K_{\text{L}} = \frac{(0.5)^2\,(25.0)}{(0.1)\,/\,(60)} = \frac{300}{D^2}$$

where the numerator is the angular acceleration at the specified condition $\omega^2 r$ and the numerator is the maximum error allowed. Checking units we have:

$$K_{\text{L}} = \frac{\left(\text{r/s}^2\right)(\text{deg})}{(\text{min})\,(\text{deg}\,/\,\text{min})} = \text{r/s}^2$$

Element Transfer Functions

The hydrostatic transmission and load can be represented by an integrator in series with a second-order system. The integration function means that the hydraulic pump displacement is proportional to the speed of the hydraulic motor, i.e. if the displacement is not in the null position, the output will continue to rotate. The second-order term is associated with the compliance of the hydrostatic drive and the inertia of the transmission and load.

The transfer function for this element was given in the specification as:

$$\frac{\theta_o}{x_o} = \frac{K_T}{D} \left[\frac{1}{(0.01D)^2 + 2(0.3)(0.01D) + 1} \right].$$

The gain K_T has units of degrees rotation per second per inch of displacement. The second-order term is expressed in the standard form showing an undamped natural frequency of $0.01^{-1} = 100.0$ radians per second and a damping ratio of 0.3. The displacement servo was specified as being a first-order lag with a time constant of 0.005 second, i.e.

$$\frac{x_o}{x_i} = \frac{1}{(1 + 0.005D)}.$$

The control algorithm of the electronic controller can be represented by the expression $\frac{K_G}{D}[C(D)]$ which is an integrator, $\frac{K_G}{D}$ in series with a dynamic compensation element $C(D)$. The integrator is necessary to provide class 2 system action which requires two integrators in the open loop transfer function.

The object of this design study is to determine what the transfer function of the compensation element should be in order to meet the specified performance and stability objectives. The system block diagram (see Figure 3.34) can now be constructed.

From the requirements evaluation we know that the product of all the gains around the loop must equal $300.0\,\mathrm{sec}^{-2}$, i.e.

$$K_L K_T = 300.0$$

We can now plot the uncompensated open loop response which is presented as the Bode diagram of Figure 3.35. As indicated in the figure the system is instable with negative margins. We can also check the

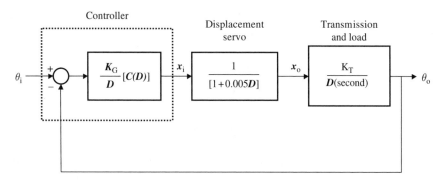

Figure 3.34 Control system block diagram

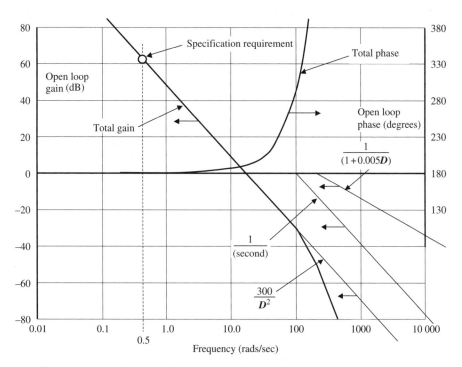

Figure 3.35 Bode diagram of the uncompensated system

dynamic performance requirement at 0.5 radians per second where the gain must be defined by the requirements per the equation:

$$\frac{\theta_O}{error} = \frac{(25.0)}{(1.0)/(60)} = 1500 = 63.52\,dB$$

This requirement is confirmed by the open loop gain line at 0.5 radians per second. We must be aware therefore that our compensation strategy to be developed does not compromise this point on the open loop gain.

As shown by the Bode diagram, the double integrator has a slope of −40 dB per decade, crossing the zero dB line at 17.3 radians per second (which is the square root of 300.0). The double integrator also imparts a phase lag of 180 degrees which is added to by the other lagging elements in the loop causing the frequency response locus to pass the 1.0, −180 degrees point on the wrong side.

To compensate this system we need to reduce the slope of the gain line to 20 dB per decade before it crosses the 0 dB line. We can do this by introducing a first-order lead element that breaks at a frequency of 10.0 radians per second.

The transfer function $(1+0.1D)/(1+0.001D)$ developed using the lead–lag approach will suffice. The denominator lag is selected to be well out of the frequency range of interest with a break frequency of 1000 radians per second and as such should not contribute significantly to the system dynamics. The new open loop transfer function is now:

$$\frac{\theta_o}{error} = \frac{300}{D^2}\left\{\frac{(1+0.1D)}{(1+0.001D)(1+0.005D)\left[(0.01D)^2+2(0.3)(0.01D)+1\right]}\right\}.$$

This gives rise to the compensated system Bode diagram of Figure 3.36 which shows the system to be stable with a phase margin of 40 degrees and a gain margin of 8 dB which is quite respectable.

We can now utilize the Nichols chart to develop the closed loop response of the system in order to examine the overshoot response as it relates to the specified limit of 40% maximum following a step change in the set point. While the frequency response does not examine the transient behavior of the system, the maximum gain at the resonant frequency can be considered as an effective guide as to the overshoot

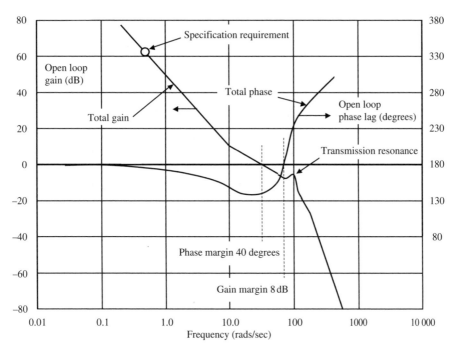

Figure 3.36 Compensated system Bode diagram

performance (for input steps that are within the linear regime of the system). Since an overshoot of 50 % is equivalent to an amplitude ratio of 1.4 or +2.9 dB we need to show that the maximum gain at the resonant frequency is less than +2.9 dB for this requirement to be met. Figure 3.37 shows the Nichols chart plot of the compensated system response which indicates a maximum closed loop gain of about +2.8 dB which is just inside the overshoot requirement.

At this point the design exercise appears to be successfully completed with the response and stability requirements met by the simple addition of a lead–lag element; however, when the system transmission was tested for the first time it was discovered that the undamped natural frequency was only half the value quoted on the design specification, i.e.

$$\frac{\theta_{\mathrm{o}}}{x_{\mathrm{o}}} = \frac{K_{\mathrm{T}}}{D}\left[\frac{1}{(0.02D)^2 + 2\,(0.3)\,(0.02D) + 1}\right].$$

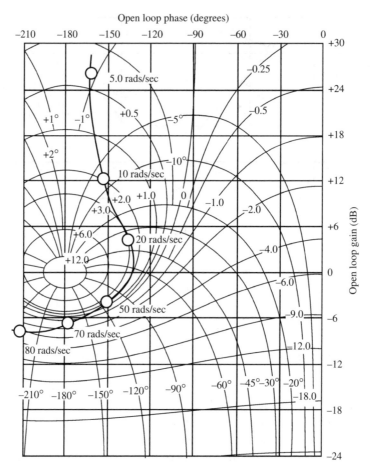

Figure 3.37 Nichols chart of the compensated system

Re-plotting the Bode diagram for the system with the actual transmission dynamics (see Figure 3.38) shows the system to be only marginally stable.

We now need to introduce additional compensation to neutralize the effects of the transmission compliance which now has a major impact on the system near to the zero dB crossover frequency. This is accomplished by introducing a second-order lead element at the same natural frequency as the transmission element together with a second-order lag with a natural frequency several times higher.

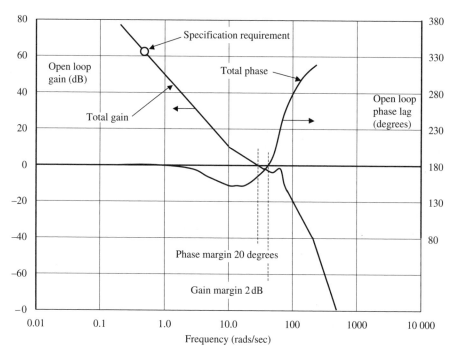

Figure 3.38 Bode diagram with actual transmission dynamics

The chosen transfer function for this compensation element is:

$$\frac{\left[(0.02D)^2 + 2(0.33)(0.02D) + 1\right]}{\left[(0.005D)^2 + 2(0.2D) + 1\right]}.$$

This element has a numerator as close as possible to the actual transmission dynamics and a denominator second-order lag having four times the transmission natural frequency. This effectively moves the transmission resonance out to 200 radians per second (32.0 Hz) which is well away from the crossover frequency. The denominator term also has a lower damping ratio that will help to further improve stability margins. It should be noted here that it is good design practice to minimize the degree of derivative action necessary in order to ensure that undesirable high frequency noise is avoided.

The revised open loop transfer function is now:

$$\frac{300}{D^2} \frac{(1+0.1D)}{(1+0.001D)(1+0.005D)\left[(0.005D)^2 + 2(0.2D) + 1\right]}.$$

Note that the numerator second-order term and the actual transmission term cancel and therefore do not contribute to the response.

Figure 3.39 shows the newly compensated system Bode diagram showing that good stability has been recovered as a result of the additional compensation elements. This figure shows that the stability margins have been improved to provide a phase margin of 58 degrees and a gain margin of about 11 dB implying well behaved closed loop response characteristics.

Finally we can plot the open loop characteristics on a Nichols chart (see Figure 3.40). From this chart we can see a significant reduction in

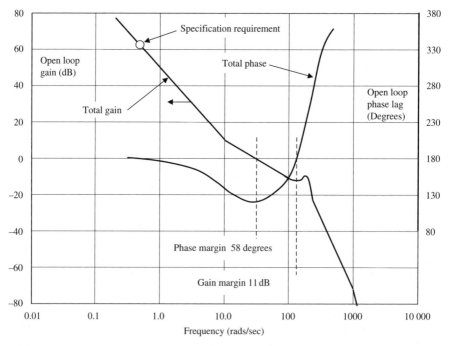

Figure 3.39 Bode diagram with revised compensation

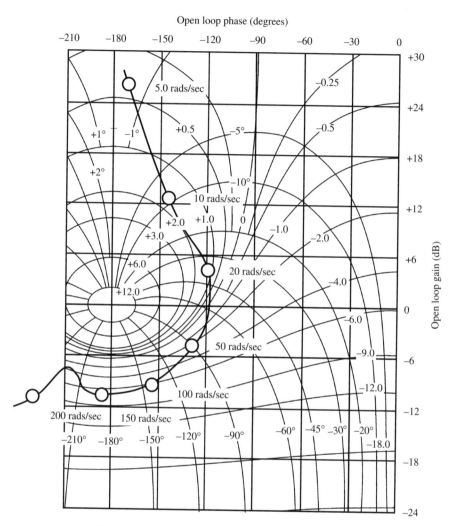

Figure 3.40 Final solution Bode diagram

the maximum closed loop gain from +2.8 dB for the original compensated solution to about +2.3 dB for the current design solution thus providing a slightly improved performance margin over the specification requirements.

From this Nichols chart we can obtain the closed loop system frequency response which is shown in Figure 3.41.

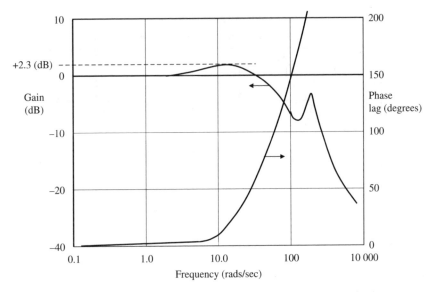

Figure 3.41   Closed loop response of the final design

As indicated in the figure the maximum gain is +2.3 dB at about 20 radians per second which represents a magnification of 1.3 or 30 % overshoot well inside the specification requirement.

An important message to be taken from this exercise is the fact that theoretical analyses typically overstate the stiffness effects leading to resonant frequencies that almost always turn out to be lower than originally predicted. In hydraulic systems the stated oil bulk modulus, which is the volumetric compliance factor, is based on pure oil with absolutely zero entrained air. This value is not typical of the empirical experience and one should be wary of this fact as a control system designer. The example presented here is not unusual in this respect and therefore it is prudent in selecting design parameters to consider what options are available should key natural frequencies turn out to be significantly lower than predicted by theoretical analyses.

3.4 Chapter Summary

Chapter 3 has built upon the basic understanding of closed loop stability analysis that was developed in Chapter 2. We learned to build compensation transfer functions from simple combinations of integrators and first-order lags that can create anticipatory elements such as:

• proportional plus integral control;
• proportional plus integral plus derivative control;
• lead–lag compensation;
• lag–lead compensation.

By selecting the appropriate values for the time constants and gains used in these building blocks the anticipatory elements can be arranged to neutralize (cancel out) one or more undesirable elements in the control loop.

An important message to the would-be practitioner is to recognize that the anticipatory (phase lead) terms contain derivatives (i.e. D terms in the numerator of transfer functions) that can be a major source of high frequency noise. It is therefore good practice to include a high frequency noise filter term when developing derivative elements. In other words, having made sure that the appropriate amount of phase lead is provided in the frequency range of interest (i.e. where the open loop gain line crosses the zero dB line) we must make sure that the derivative element is canceled out at the higher frequencies where there is no significant effect on the system being compensated.

The key to becoming skillful in the optimization of closed loop control systems is to practice the art of sketching frequency response plots, using firstly the Bode diagram and then the Nichols chart. With practice these charts can be developed quickly and the ensuing visibility provided. The ability of the Nichols chart to provide both an open and closed loop view of the system makes an important contribution to the development of a 'feel for the problem'.

4

Introduction to Laplace Transforms

This chapter introduces the 'Laplace transform' as a tool for analyzing both the transient and frequency response characteristics of control systems. Since it is used extensively by the control systems community it is important that we address the topic in this book. While the term 'transform' implies mathematical complexity the good news is that the subject can be covered at a fairly high level leaving the reader able to readily apply Laplace transforms to control system problems and to benefit substantially from the insight into the complex frequency domain that is provided. For completeness the mathematics of the Laplace transform is developed here for those readers interested in the proof of the general case; however, it is not essential reading for applying Laplace transforms to control systems analysis or for appreciating the visibility into the functional behavior of control systems that this topic provides.

4.1 An Overview of the Application of Laplace Transforms

Provided here is a brief overview of the process associated with the application of Laplace transforms in order to determine the response of linear systems to a multitude of input stimuli including step, ramp and sinusoidal inputs. In order to apply Laplace transforms to solve response problems we convert the transfer functions we have been working with

Stability and Control of Aircraft Systems: Introduction to Classical Feedback Control   R. Langton
© 2006 John Wiley & Sons, Ltd

in previous chapters from the time domain into the complex frequency domain, also referred to as the 's' plane in order to eliminate the independent variable, 't'. Thus we can say:

$F(s) = \mathcal{L}f(t)$, where the term '$\mathcal{L}$' denotes Laplace transform

and $F(s)$ is the 's' plane equivalent of $f(t)$.

Once in the 's' plane it becomes a simple algebraic task to determine the response of the system to, for example, a step input and, using the reverse process designated the 'inverse Laplace transform' ($\mathcal{L}^{-1}$), to express the answer as a function of time. We can consider this process as similar to using logarithms where we simplify the processes of multiplication, division and exponent operations by converting numbers into logarithms via tables and after performing the simplified process (e.g. adding the logarithms to generate the product) we use the 'anti-logarithm' table to convert the answer back into the real world.

In the 's' plane the frequency response of a control system can be obtained simply by substituting $s = j\omega$ in the Laplace transform of the system just as we did before with the D operator so that all the effort we have put into how to develop the frequency response of a system still applies in the 's' plane. This will be explained in detail in the paragraphs that follow.

4.2 The Evolution of the Laplace Transform

So what is the Laplace transform, why is it needed and what is the implication of the 's' plane or complex frequency domain? We will try to address these questions here without getting into any complex mathematics. Let us begin by going back to the application of the 'D' notation in the expression of differential equations and transfer functions. To begin we shall adopt a more specific definition of the derivative with respect to time, d/dt. Consider:

$$s = \frac{d}{dt} \text{ and } \frac{1}{s} = \int_0^t (\ ) \, dt.$$

This appears to be the same definition used for the D operator; however, because we have applied specific limits to the integration process above we say that this is a *definite* integral. So we can say:

$$st = \frac{\mathrm{d}\,(t)}{\mathrm{d}t} = 1$$

and

$$\frac{1}{s}\,(t) = \int_0^t (t)\mathrm{d}t = \left[\frac{t^2}{2}\right]_0^t = \frac{t^2}{2}$$

which is also *definite*.

We now need to introduce the concept of the 'unit impulse' which is a very narrow, very tall pulse of *unit area* occurring just after $t = 0$. This is perhaps the most important concept in the use of Laplace transforms as we shall see later. The unit impulse is defined as δt and is shown graphically in Figure 4.1. It can be seen from the figure that $\delta t = 0$ so long as $t \neq 0$.

We can also see that if we integrate the impulse function between $t = 0$ and any positive value of t, the answer will always be 1.0 because the area under the impulse function is, by definition, ***unit area***.

Therefore we can say mathematically:

$$\int_0^t (\delta t)\mathrm{d}t = 0 \text{ for } t < 0 \text{ and } 1 \text{ for } t > 0.$$

We can represent this function graphically as shown in Figure 4.2 which clearly depicts the step function denoted by $H\,(t)$. Using the

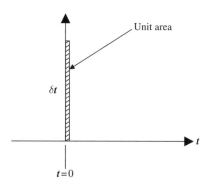

Figure 4.1   The unit impluse function

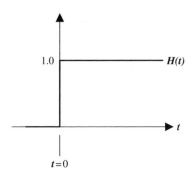

Figure 4.2 The step function $H(t)$

same principles developed by the use of the 'D' notation concept we can say:

$$\frac{1}{s}(\delta t) = H(t).$$

Similarly we can say:

$$\frac{1}{s^2}(\delta t) = 0 \text{ for } t < 0 \text{ and } t \text{ for } t > 0.$$

Continuing the same logic:

$$\frac{2}{s^3}(\delta t) = 0 \text{ for } t < 0 \text{ and } t^2 \text{ for } t > 0.$$

From the above exercise we have found the 's' plane transfer functions for the 'boxes' whose response to a unit impulse is a step, a ramp and a parabola. Figure 4.3 illustrates this concept in both block diagram and graphical form.

These 's' plane transfer functions are in fact the Laplace transforms for the step, ramp and parabola time functions. What we need now is to develop the formula that allows us to define what goes in the 'box' for any function of time whose input is the impulse function δt that will give the desired time response for $t > 0$.

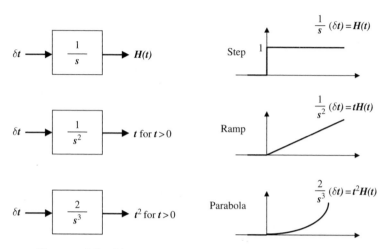

Figure 4.3 Step, ramp and parabola transforms

The following formula defines the general case for the transformation of any time function into the 's' plane

$$\mathcal{L}f(t) = F(s) = \int_0^\infty f(t)e^{-st}dt.$$

Let us now check the above transform definition using the step and ramp functions previously determined

$$\mathcal{L}H(t) = \int_0^\infty (1)e^{-st}dt = \frac{1}{s}$$

$$\mathcal{L}(t) = \int_0^\infty te^{-st}dt = \frac{1}{s^2}.$$

This supports the arguments developed above and Table 4.1 shows a number of Laplace transforms for common time functions.

4.2.1 Proof of the General Case

This section provides the proof of the general case for transforming any function of time into the 's' plane. As mentioned in the introduction to

Table 4.1 Commonly used Laplace transforms

Function	$f(t)$	$F(s)$	Block diagram
	$\delta(t)$	1	$\delta(t) \rightarrow \boxed{1} \rightarrow \delta(t)$
	$H(t)$	$\frac{1}{s}$	$\delta(t) \rightarrow \boxed{\frac{1}{s}} \rightarrow H(t)$
	t	$\frac{1}{s^2}$	$\delta(t) \rightarrow \boxed{\frac{1}{s^2}} \rightarrow tH(t)$
	$e-kt$	$\frac{1}{s+k}$	$\delta(t) \rightarrow \boxed{\frac{1}{s+k}} \rightarrow e^{-kt}H(t)$
	$\sin \omega t$	$\frac{\omega}{s^2+\omega^2}$	$\delta(t) \rightarrow \boxed{\frac{\omega}{s^2+\omega^2}} \rightarrow \sin \omega tH(t)$
	$\cos \omega t$	$\frac{s}{s^2+\omega^2}$	$\delta(t) \rightarrow \boxed{\frac{s}{s^2+\omega^2}} \rightarrow \cos \omega tH(t)$

this chapter it is presented here for completeness and is not considered as essential reading for those interested in the application of the Laplace transform to linear control systems analysis. Those readers who do not feel the need to cover the detailed mathematical development of the Laplace transform, should go to Section 4.3.

To begin we must first recall Taylor's Theorem:

$$f(x+h) = f(x) + hf'(x) + \frac{h^2 f''(x)}{2!} + \frac{h^3 f'''(x)}{3!} + \ldots\ldots$$

Introducing the definition: $s = \mathrm{d}/\mathrm{d}t$ we obtain:

$$f(x+h) = f(x) + hs(x) + \frac{h^2 s^2(x)}{2!} + \frac{h^3 s^3(x)}{3!} + \ldots.$$

$$= f(x)\left(1 + hs + \frac{h^2 s^2}{2!} + \frac{h^3 s^3}{3!} + \ldots.\right).$$

Thus we can write:

$$f(x+h) = e^{hs}f(x).$$

and therefore:

$$f(x-h) = e^{-hs}f(x).$$

Consider now a unit impulse occurring at some time τ rather than at $t = 0$, we would express this function in our Laplace transform terminology as

$$\delta(t - \tau) = e^{-st}\delta(t) \tag{4.1}$$

If we now consider the integral:

$$\int_0^\infty f(\tau)\delta(t - \tau)\,d\tau,$$

referring to Figure 4.4 we can see that since $\delta(t - \tau) = 0$, if $(t - \tau) \neq 0$, the only significant value of $f(\tau)$ is $f(t)$ and this is constant with respect to τ. Therefore

$$\int_0^\infty f(\tau)\delta(t - \tau)\,d\tau = f(t)\int_0^\infty \delta(t - \tau)\,d\tau = f(t)(1).$$

Hence

$$f(t) = \int_0^\infty f(\tau)\delta(t - \tau)\,d\tau \tag{4.2}$$

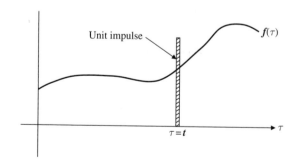

Figure 4.4 Function $f(\tau)$ is significant only at $\tau = t$

If we now put (4.1) in (4.2) we have

$$f(t) = \int_0^\infty f(\tau) e^{-st} \delta(t) \, d\tau$$

and since $\delta(t)$ is constant with respect to τ we can write

$$f(t) = \left(\int_0^\infty f(\tau) e^{-st} d\tau \right) \delta(t) \qquad (4.3)$$

If we now denote the integral in equation (4.3) as $F(s)$ and replace τ by t in the integral since it is a dummy variable, we arrive at the equation below and its associated block diagram of Figure 4.5:

$$f(t) = F(s) \delta(t).$$

We have thus defined the integral that transforms any function of time into the 's' domain as:

$$F(s) = \int_0^\infty f(t) e^{-st} dt$$

which is defined as the Laplace transform of $f(t)$.

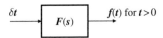

Figure 4.5 Block diagram of the Laplace transforms process

4.3 Applying Laplace Transforms to Linear Systems Analysis

The next step is to relate all of this Laplace transforms material to differential equations and linear control systems analysis. Consider now the Laplace transform of the time derivative of a time function $f(t)$, i.e

$$\mathcal{L} \frac{d(f(t))}{dt} = \int_0^\infty \left[\frac{d(f(t))}{dt} \right] e^{-st} dt.$$

By integrating the right-hand side of the above equation by parts we obtain:

$$\left[f\left(t\right)e^{-st}\right]_0^\infty + s\int_0^\infty \left[f\left(t\right)\right]e^{-st}\mathrm{d}t$$
$$= f\left(t\right)_0 + s\mathcal{L}f\left(t\right)$$

where $f\left(t\right)_0$ is the value of $f\left(t\right)$ at $t = 0$. Therefore if we can assume that $f\left(t\right) = 0$ at $t = 0$ (and this is reasonable for most control systems work) then

$$\mathcal{L}Df\left(t\right) = s\,\mathcal{L}f\left(t\right).$$

This states that the Laplace transform of the derivative of a time function is simply 's' multiplied by the function itself (provided that the initial conditions are zero at $t = 0$). Thus we can simply replace the operator D in our traditional transfer functions by the Laplace operator 's' to obtain the Laplace transform equivalent of that transfer function. Let us now go back to our original time domain, D operator depiction of a control system as shown in Figure 4.6. If the initial conditions at $t = 0$ are zero we can replace $f\left(D\right)$ by $F\left(s\right)$ based on what we have learned above.

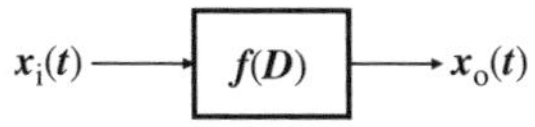

$$x_i(t) \longrightarrow \boxed{f(D)} \longrightarrow x_o(t)$$

Figure 4.6 Typical time domain transfer function depiction

So we can say:

$$x_o\left(t\right) = f\left(D\right)x_i\left(t\right)$$

which is the same as

$$x_o\left(t\right) = F\left(s\right)x_i\left(s\right)\delta\left(t\right)$$

where $x_i\left(s\right)$ is the Laplace transform of $x_i\left(t\right)$ and δt makes sure that the output remains at zero until the impulse arrives at $t = 0$.
Similarly we can say:

$$x_o\left(t\right) = x_o\left(s\right)\delta\left(t\right).$$

If we now express everything in the 's' domain we can say:

$$x_o(s)\,\delta(t) = F(s)\,x_i(s)\,\delta(t)\,.$$

Since the unit impulse functions cancel we have a new 's' domain block diagram as shown in Figure 4.7. Everything is now in the 's' domain and the independent variable t has been eliminated. All we have to do to derive the response of any system to an input disturbance is to convert the time variable disturbance into the 's' domain by finding its Laplace transform (from tables).

Figure 4.7 's' domain transfer function depiction

The algebraic product of the input and system Laplace transforms defines the response in the 's' domain. If we now take the inverse transform we obtain the time domain response. Let us now demonstrate this process using a simple first-order lag as the system and a step input as the input disturbance. Figure 4.8 shows this problem statement in both the time domain and the 's' domain.

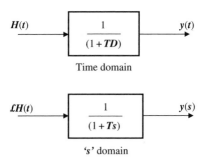

Figure 4.8 First-order lag shown in time and 's' domains

Since we know that the Laplace transform of the step function, $\mathcal{L}H(t) = 1/s$, we can say:

$$y(s) = \frac{1}{s}\frac{1}{(1+Ts)}\,. \tag{4.4}$$

Using partial fractions (remember them?) we can express the right-hand side of this equation as the sum of two separate terms as follows:

$$y\left(s\right) = \frac{1}{s} - \frac{T}{\left(1+Ts\right)} = \frac{1}{s} - \frac{1}{\left[s+\left(1/T\right)\right]}.$$

If we refer to the table of Laplace transforms in Table 4.1 we can convert each term into the time domain to obtain:

$$y\left(t\right) = 1 - e^{-t/T} \text{ for } t \geq 0$$
$$= \left(1 - e^{t/T}\right) H\left(t\right).$$

This is solution is illustrated graphically in Figure 4.9. We have just used Laplace transforms to calculate the step response of a first-order lag.

Let us now look at what all this means in the 's' plane (or domain). Figure 4.10 shows the 's' plane where 's' is represented by the complex number $\sigma + j\omega$. This can be considered as the 'complex frequency' where the imaginary term is pure frequency and the real term determines the degree of exponential decay (or growth) and for this reason the 's' plane is usually referred to as the 'complex frequency domain'.

The boxes in the figure located in various positions around the plane show the equivalent time responses to a unit impulse at $t = 0$. For points in the plane that lie on the imaginary axis, the real term is zero and therefore the exponential decay/growth is zero and the result is

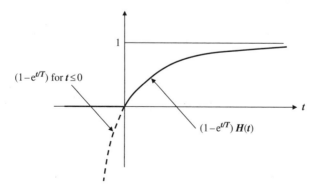

Figure 4.9 Step response of a first-order lag

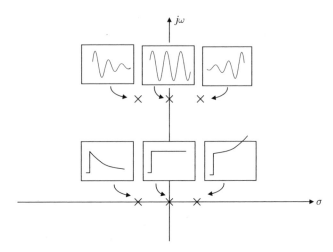

Figure 4.10 Time response implications of the 's' plane

pure frequency. Points that lie on the real axis imply zero frequency (i.e. $j\omega = 0$) and there will be no oscillations. For points in the plane on the left-hand side of the imaginary axis (negative values of σ) the rate of exponential decay increases as the value of σ becomes increasingly negative. The opposite is true for points in the plane to the right of the imaginary axis. Here the exponential term implies growing oscillations with the growth rate increasing for increasing values of σ. The negative $j\omega$ half of the 's' plane is simply a mirror image of the positive half of the plane which is inherent in the mathematics and this mirror image represents an important feature in control systems in that when imaginary roots occur (implying oscillatory behavior) they always occur in complex conjugate pairs of the form $\sigma \pm j\omega$. This complex frequency interpretation of points in the 's' plane provides a valuable insight into the effect that each of the roots of the control system transfer function have on the behavior of the system in the real world.

Going back to our first-order lag step response problem, this can be represented by the 's' plane diagram of Figure 4.11. The $\times$'s are defined as 'poles' meaning that for those specific values of 's' the function $y(s)$ defined by equation (4.4) above becomes infinite hence the term 'pole'. As indicated in Figure 4.11 the pole at $-1/T$ is referred to as the 'system pole' and the pole at the origin as the 'input pole' representing the step function.

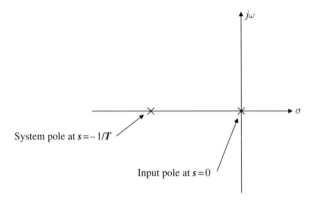

Figure 4.11 's' plane representation of the first-order lag response

Extending this to the general case we can say that linear systems are typically of the form:

$$F(s) = k\frac{\left(s^m + a_1 s^{(m-1)} + a_2 s^{(m-2)} + \ldots a_m\right)}{\left(s^n + b_1 s^{(n-1)} + b_2 s^{(n-2)} + \ldots b_n\right)}.$$

Factorizing we obtain:

$$F(s) = \frac{k\left(s - z_1\right)\left(s - z_2\right) \ldots \left(s - z_m\right)}{\left(s - p_1\right)\left(s - p_2\right) \ldots \left(s - p_n\right)}.$$

The roots of the numerator $z_1, z_2 \ldots z_m$ are called the 'zeros' of the system since when $s = z$ the term equals zero and the function $F(s)$ goes to zero. The roots of the denominator $p_1, p_2 \ldots p_n$ are the poles as described in the first-order lag example above so that when $s = p$ the function $F(s)$ goes to infinity. To illustrate the 's' plane pole–zero concept graphically we can represent the magnitude of $F(s)$ (i.e. the modulus$|F(s)|$) as a surface in the 's' plane as illustrated by Figure 4.12.

4.3.1 Partial Fractions

In view of the importance of partial fractions as the process used to convert Laplace transforms back from the 's' domain into the time domain, the subject is revisited in more detail in this section. Let us begin

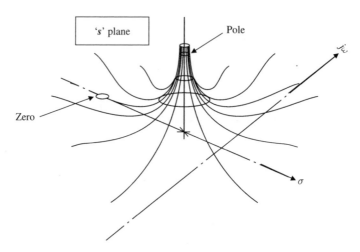

Figure 4.12 Graphical representation of **F(s)**

by considering the general case for linear systems in the 's' domain. Partial fractions allow us to express the general equation:

$$F(s) = \frac{k(s - z_1)(s - z_2) \dots (s - z_m)}{(s - p_1)(s - p_2) \dots (s - p_n)}$$

as the sum of a number of separate fractions of the form:

$$F(s) = \frac{A_1}{(s - p_1)} + \frac{A_2}{(s - p_2)} + \dots \frac{A_n}{(s - p_n)}.$$

This rule is provided that the number of poles is greater than the number of zeros and also that there are no repeated poles. (If repeated poles do occur it is easy to circumvent this limitation by moving the location of one of the repeated poles slightly.)

The A values in the numerators are called the 'residues' at the poles and the method for calculating the residues is described in the following example. To calculate the value of the residue A_r at pole p_r we simply set $s = p_r$ in the factorized equation for $F(s)$, and cover up (ignore) the factor $(s - p_r)$. That is

$$A_r = \frac{k (p_r - z_1) (p_r - z_2) \ldots (p_r - z_m)}{(p_r - p_1) (p_r - p_2) \ldots (****) \ldots (p_r - p_n)}.$$

The term $(****)$ in the above expression represents the factor $(s - p_r)$ which is ignored in the calculation for A_r. What this means is that the residue A_r at pole p_r is defined as

$$A_r = \frac{(\text{product of vectors from zeros to pole at } p_r)}{(\text{product of vectors from poles to pole at } p_r)}.$$

The residue at a pole, therefore, is a vector quantity so that: $A_r = |A_r| \angle A_r$.
To reinforce this process let us consider the example shown in Figure 4.13 which shows a system block diagram and the equivalent 's' plane representation.

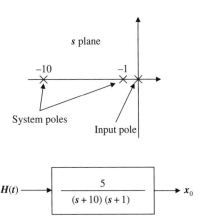

Figure 4.13 Block diagram and 's' plane representation

The output $x_o(s)$ is the product of the input and system Laplace transforms, i.e.

$$x_o(s) = \frac{1}{s} \left[\frac{5}{(s+10)(s+1)} \right].$$

In partial fraction form we obtain:

$$X_o(s) = \frac{A_{-10}}{(s+10)} + \frac{A_{-1}}{(s+1)} + \frac{A_0}{s}.$$

Using the previously described method we can calculate the residues:

$$A_{-10} = \frac{(5)(1)}{(-10)(-9)} = \frac{5}{90}; \quad A_{-1} = \frac{(5)(1)}{(+9)(-1)} = -\frac{5}{9}; \quad A_0 = \frac{(5)(1)}{(10)(1)} = \frac{5}{10}.$$

Instead of applying the partial fractions 'formula' for obtaining the residues at each pole we can use the vector product approach as shown in Figure 4.14 to calculate the residues. As indicated in the figure this yields the same values as before. In this particular example the calculations are simple since there are no zeros and all of the poles lie on the real axis.

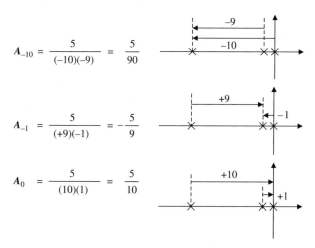

Figure 4.14 Vector approach to residue calculation (in each case the loop gain **k**=5)

Therefore the vector angles are either zero or 180 degrees. Inserting the values for the residues into the response equation we obtain the 's' plane expression:

$$x_o = \frac{5}{90}\left[\frac{9}{s} - \frac{10}{(s+1)} + \frac{1}{(s+10)}\right]$$

From the transforms table (Table 4.1) we can now express this function as a function of time:

$$x_o(t) = \frac{5}{90}\left(9 - 10e^{-t} + e^{-10t}\right).$$

This function is shown in Figure 4.15 where it can be seen that the contribution from the pole at $s = -10$ is small compared with the other terms. This is an important observation since it means that a pole isolated from the main group of poles will have a small residue and contribute little to the overall response.

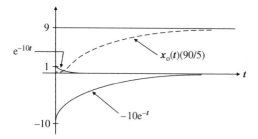

Figure 4.15   Time response plot

To complete this segment we need to examine the residues associated with oscillatory roots, i.e. roots that are not on the real axis of the 's' plane. First let's connect the 's' plane representation of an oscillatory second-order system to the standard form of second-order transfer function developed back in Chapter 1 which is:

$$\frac{1}{\dfrac{s^2}{(\omega_n)^2} + \dfrac{2\zeta s}{\omega_n} + 1}.$$

Here ω_n is the undamped natural frequency and ζ is the damping ratio. Multiplying the numerator and denominator by $(\omega_n)^2$ gives the following expression which is now in the 's' plane format, i.e.

$$\frac{(\omega_n)^2}{s^2 + 2\zeta\omega_n s + (\omega_n)^2}.$$

We can obtain the roots of the numerator using the old high school formula:

$$\frac{-b \pm \sqrt{b^2 - 4ac}}{2a}.$$

From this it is clear that when the square root term becomes negative the roots must always be complex conjugates. Substituting the standard second-order expression coefficients in the root finding equation yields:

$$\frac{-2\zeta\omega_n \pm \sqrt{(2\zeta\omega_n)^2 - 4(\omega_n)^2}}{2} = -\zeta\omega_n \pm \omega_n\sqrt{(\zeta^2 - 1)}.$$

Considering values of $\zeta < 1$, the second term will always be imaginary and referring to the 's' graph of Figure 4.16 we can see that the real coordinate, $-\omega_n \cos\theta$, must equate to $-\zeta\omega_n$, therefore the damping ratio $\zeta = \cos\theta$. Similarly, the imaginary coordinate $\pm\omega_n \sin\theta$ must equate to $\pm\omega_n\sqrt{(\zeta^2 - 1)}$. Substituting $\zeta = \cos\theta$, the term inside the square root sign becomes $\sin^2\theta$ thus confirming the location of the complex conjugate roots as $(\omega_n)\cos\theta \pm j(\omega_n)\sin\theta$.

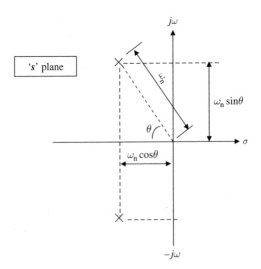

Figure 4.16 Definition of natural frequency and damping ratio

We have now related the standard expression for a second-order system which is defined in terms of the undamped natural frequency and damping ratio to the graphical representation of two complex conjugate poles in the 's' plane. As indicated earlier, oscillatory roots always come in 'complex conjugate pairs' for example:

$$\frac{10}{(s^2 + 2s + 10)} = \frac{10}{[s + (1 + 3j)][s + (1 - 3j)]}.$$

In the 's' plane these complex roots are located as shown in Figure 4.17 where the imaginary coordinates $\pm 3j$ determine the frequency of oscillation and the real term, -1, determines the rate of decay of the oscillations.

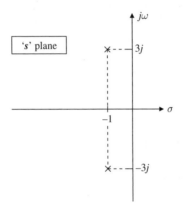

Figure 4.17 Second-order system poles

In the special case where there is zero damping, i.e. where the σ term is zero, we will have sustained oscillations with the poles lying on the imaginary axis as shown in Figure 4.18.

In this case the system is defined by the transfer function:

$$\frac{5}{(s^2 + 25)} = \frac{5}{(s + 5j)(s - 5j)}$$

Note here the similarity of the Laplace transform of $\sin(\omega t) = \frac{\omega}{(s^2 + \omega^2)}$ in the table of commonly used Laplace transforms (Table 4.1). From this we can see that when $s = j\omega$ we get pure frequency which ties in with

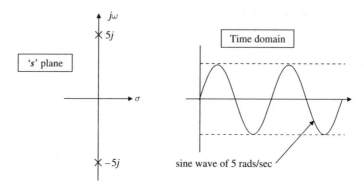

Figure 4.18   Second-order system poles with zero damping

the frequency response analysis method where we substitute $s = j\omega$ (or $D = j\omega$) in the transfer function. Residues at complex conjugate poles typically take the form $\frac{A}{2j}$ and $\frac{A^*}{-2j}$ where the residues A and A^* are conjugates.

Therefore the general expression for the impulse response of a complex conjugate pair will be of the form:

$$F(s) = \frac{1}{2j}A\left\{\frac{1}{[s + (\sigma - j\omega)]}\right\} - \frac{1}{2j}A^*\left\{\frac{1}{[s + (\sigma + j\omega)]}\right\}.$$

The corresponding t terms are:

$$f(t) = \frac{1}{2j}\left(Ae^{-(\sigma - j\omega)} - \text{conjugate}\right)$$

$$= \frac{1}{2j}\left(|A|\,e^{-\sigma t}e^{j(\omega t + \angle A)} - \text{conjugate}\right)$$

and since we know that

$$\sin(\ ) = \frac{1}{2j}\left(e^{(\ )} - e^{-(\ )}\right)$$

we can write:

$$f(t) = |A|\,e^{\sigma t}\sin(\omega t + \angle A).$$

Therefore we can say that the impulse response of this system is a damped sinusoid with a time constant of $1/\sigma$ seconds and a frequency of ω radians per second.

Since Laplace transforms provide a complete solution to linear differential equations and their stimulus we can obtain both the transient response to any function of time (provided that we can define the input time function as a Laplace transform) and also the frequency response by substituting $s = j\omega$ into the system transfer function. Suppose we have a system defined by the Laplace transform:

$$F(s) = \frac{k(s-z_1)(s-z_2)\ldots(s-z_m)}{(s-p_1)(s-p_2)\ldots(s-p_n)}.$$

At any point q in the 's' plane we can determine the value of $F(s)$ by the previously described expression which means:

$$F(s) = \frac{k(\text{product of vectors from the zeros to point } q)}{(\text{product of vectors from the poles to point } q)}.$$

This point is illustrated graphically by Figure 4.19 which shows an arbitrary point q in the 's' plane with the vectors from each zero and pole in a typical system. The vector lengths are multiplied together and the vector angles summed to obtain the vector value of $F(s)$ at the selected point q.

Suppose now that we select point q to be on the imaginary axis of the 's' plane. Here $s = j\omega$ and so $F(s)$ will describe the frequency response of the system at that specific frequency. Thus moving along the $j\omega$ axis and calculating the ratio of the vector products in the above expression

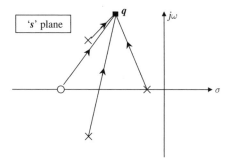

Figure 4.19 Vectors to point 'q'

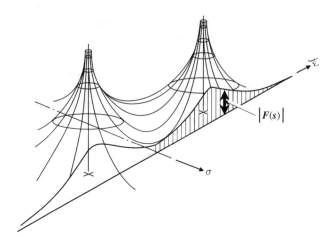

Figure 4.20 Interpretation of a second-order system frequency response

for $F(s)$ generates the frequency response of the system defined in the 's' plane. Once again we can visualize the amplitude ratio (i.e. the modulus of the function $F(s)$) as a three-dimentional surface in the 's' plane as shown in Figure 4.20. As indicated in the figure, the amplitude ratio of the frequency response can be interpreted as a section through the imaginary ($j\omega$) axis. In this second order system example there are two complex conjugate poles fairly close to the $j\omega$ axis. Therefore a section through the amplitude surface of the 's' plane along the $j\omega$ axis will result in a cut that penetrates well into the side of the 'mountain' associated with the poles. For this particular system the frequency response will show a significant magnification, or resonance, at a frequency adjacent to the poles as indicated by the shape of the section shown in the figure.

4.4 Laplace Transforms – Summary of Key Points

Laplace transforms allow us to operate in the 's' plane giving us a unique insight into the contributions that each element in the control loop makes to the overall response. Laplace transforms for almost any useful time functions can be obtained from tables (just like using logarithms). Once in the 's' domain, it becomes easy to generate mathematical solutions for the response of a system by obtaining the product of the stimulus and the system Laplace transforms and, with the help of partial fractions, converting the result back into the time domain.

Here are some of the major points established regarding Laplace transforms:

(1) The Laplace transform for any function of time can be determined from the expression:

$$\mathcal{L}f(t) = F(s) = \int_0^\infty f(t)e^{-st}dt.$$

Laplace transforms of all commonly used time functions are available from standard tables.

(2) The Laplace transform of a derivative of a function of time, i.e.

$$\mathcal{L}\frac{d}{dt}f(t) = sF(s)$$

and similarly:

$$\mathcal{L}\frac{d}{dt^2}f(t) = s^2F(s)$$

provided that the initial conditions are zero at $t = 0$. This allows allows us to convert our traditional transfer functions into the 's' domain by simply substituting s for the operator D in our block diagrams.

(3) An important observation regarding Laplace transforms is that the time response of a function $F(s)$ is the response of that system to a unit impulse $\delta(t)$.

(4) The frequency response of a system defined by the Laplace transform $F(s)$ can be obtained by substituting $s = j\omega$ just as we did with the transfer functions using the D operator.

(5) When converting system responses from the 's' domain back into the time domain we use partial fractions to define the separate contributions made to the solution by each pole. In this form it is easy to use the standard tables to establish the composite time response.

(6) The contribution of each pole to the overall response is defined by the residue at that pole. Poles located away from the main group will have small residues and contribute little to the response.

4.5 Root Locus

It would be irresponsible not to mention the root locus method of linear control system design and analysis because it is so closely linked to Laplace transforms. We introduce this topic here, therefore, to provide the reader with an interesting and insightful approach to system design through the process of optimizing the loop gain. Root locus involves working in the 's' plane and therefore the system transfer functions are defined by their Laplace transforms.

To begin, let us revisit the baseline definition of the system closed loop transfer function (CLTF):

$$\frac{\text{output}}{\text{input}} = \frac{(\text{forward path})}{(1 + \text{loop})}.$$

The denominator, when set equal to zero, is called the 'characteristic equation' of the system. This equation's solution defines the transient response characteristics of the system; however, it is important to realize that the definition of the specific locations of 'input' and 'output' around the loop do not affect the definition of the characteristic equation as indicated graphically by Figure 4.21. Each of the block diagrams in

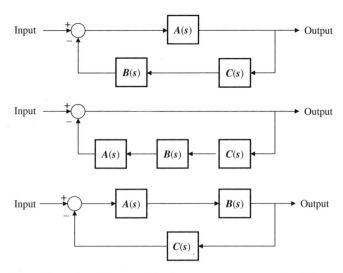

Figure 4.21 Three systems with the same characteristic equation

Figure 4.21 will have output responses to input stimuli that are different; however, the inherent stability of each system will be the same because they have the same characteristic equation, i.e.

$$1 + A(s) B(s) C(s) = 0.$$

A typical system characteristic equation may have the following definition:

$$1 + \frac{k(s+z_1)(s+z_2)}{(s+p_1)(s+p_2)(s+p_3)} = 0$$

where z_1 and z_2 are zeros in the open loop transfer function and p_1, p_2 and p_3 are poles. The term k is a measure of the product of all of the gains around the loop.

Rearranging the characteristic equation, we can say:

$$\frac{k(s+z_1)(s+z_2)}{(s+p_1)(s+p_2)(s+p_3)} = -1 = |1.0| \angle 180^{o}.$$

It follows that if we can define a locus of all of the points in the 's' plane in which the sum of all the zero angles minus the sum of all the pole angles is always 180 degrees, we will be able to see immediately how the 'closed loop roots' of the system move as k varies from zero to infinity. The value of k at any point along the locus is simply the product of the vector lengths (moduli) from the zeros to the point divided by the product of the vector lengths from the poles to that point. This is difficult to conceive at first but via the use of examples all will become clear. Meanwhile let us learn how to construct these root loci.

4.5.1 Root Locus Construction Rules

Root loci can be constructed using a few simple rules which are easy to apply to typical linear control systems. These rules are developed below followed by an number of examples to illustrate how the application of root locus theory can provide a clear picture of first, how the loop gain affects closed system stability, and second, how the system can be best modified through the addition of compensating poles and zeros to optimize closed loop performance.

Rule 1   Root loci travel from the poles (where $k = 0$) to the zeros (where $k = \infty$). Figure 4.22 shows a graphical depiction of two loci traveling from one pair of poles to another pair of zeros.

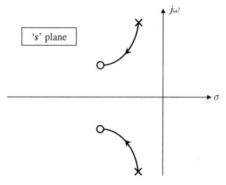

Figure 4.22   Root loci travel from poles to zeros

Rule 2   A locus will always be present along the real axis to the left of a number of odd poles plus zeros (see Figure 4.23).

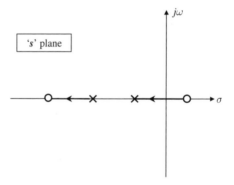

Figure 4.23   Root loci with real axis poles and zeros

Rule 3   When there are more poles than zeros (most cases), additional zeros are located at infinity. These loci will be asymptotic to straight lines with angles defined by the equation:

$$\theta = \frac{180 + i360}{n - m}$$

where m and n are the number of zeros and poles respectively.

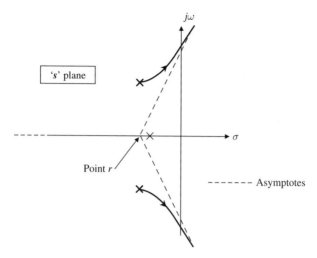

Figure 4.24 Root loci asymptotes for a three pole system

This equation applies for $i = 0, \pm 1, \pm 2$, etc. until all $(n-m)$ angles not differing by multiples of 360 degrees are obtained. Figure 4.24 shows the loci for a system of three poles and no zeros illustrating the asymptotes for $\theta = 60°, 180°, 300°$.

Rule 4   The starting point on the real axis from which the asymptotes radiate (point r in Figure 4.24) is given by the equation:

$$\vartheta = \frac{\sum \text{pole\_values} - \sum \text{zero\_values}}{n - m}.$$

This point is the centroid of the asymptotes. Figure 4.25 illustrates this rule for a system with two real poles at -2.0 and -9.0, two complex conjugate poles at $-4.0 \pm 4j$ and a zero at -5.0. Inserting the values into the above equation gives:

$$\vartheta = \frac{(-9 - 4 + 4j - 4 - 4j - 2) - (-5)}{4 - 1} = -4.67.$$

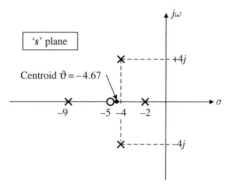

Figure 4.25   Asymptote centroid for a specific system

Figure 4.26 shows the root loci for this system. On the real axis the locus from the pole at −2.0 goes to the zero at −5.0 and the locus from the pole at −9.0 goes to the zero at minus infinity. The complex poles merge with the asymptote that radiates from the real axis at −4.67. Any point on any of the loci represents a specific value of *k* and a corresponding point on each of the other loci will represent that same value of *k*. These points on the loci having the same

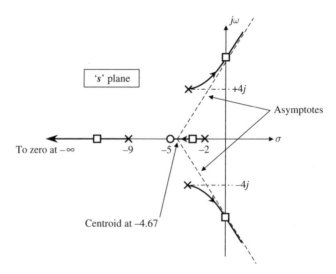

Figure 4.26   Root locus example showing closed loop roots

value of **k** are roots of the characteristic equation and, as such, are roots of the closed loop system. In Figure 4.26 the small rectangles on each locus represent the specific value of **k** for the system to be marginally stable. This is because the roots along the loci from each of the complex poles are just crossing the imaginary axis and would therefore exhibit sustained oscillations at about 8 radians per second. The other roots along the real axis loci correspond to that same value of loop gain.

Rule 5 When two adjacent poles lie on the real axis, there will be a breakaway point on the locus between these two poles. These breakaway points are defined:

- the locus leaves the real axis at the maximum possible value of **k** in that region;
- the locus joins the real axis at the minimum possible value of **k** in that region.

Figure 4.27 illustrates this rule.

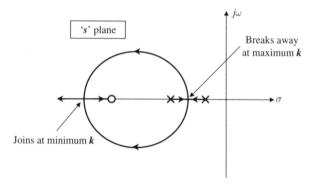

Figure 4.27 Breakaway point example

Rule 6 The angle of departure of a locus from a complex pole can be determined from the following equation, where ϕ is the angle of departure:

$$\phi = 180 - \sum (p) + \sum (z)$$

where $\sum (p) = \sum$ (pole angles to the pole), and $\sum (z) = \sum$ (zero angles to the pole).

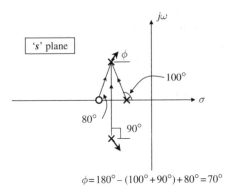

$$\phi = 180° - (100° + 90°) + 80° = 70°$$

Figure 4.28 Angle of departure example

Figure 4.28 illustrates this rule for a system with two complex conjugate poles, a real pole and a real zero.

The above six rules of root locus construction make it easy to generate the root loci for any linear system and to visualize how the closed loop roots move as the loop gain is varied from zero to infinity.

4.5.2 Connecting Root Locus to Conventional Linear Analysis

Before applying these rules to a specific control system example we need to have a feel for how root locus and the 's' plane relate to the fundamentals of control systems analysis already covered.

To begin, let us consider a first-order control loop defined by a single integrator with gain K in the forward path and a unity feedback closing the loop (see Figure 4.29). The characteristic equation for this simple system is:

$$1 + \frac{K}{s} = 0.$$

This can be represented in the 's' plane by a single pole at the origin as shown in Figure 4.29. The root locus for this control loop is a single locus that simply runs along the real (σ) axis from the pole at the origin towards minus infinity. Selecting a value of the gain K for this example determines the location of the one and only closed loop root as indicated

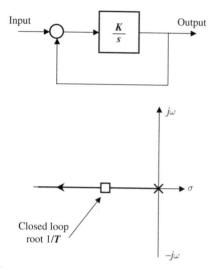

Figure 4.29 Simple first-order example

by the rectangle on the locus shown in the figure. Increasing the value of **K** simply moves the root along the σ-axis to a more negative location.

When considering the 's' plane, we observe that any point in the plane defined by $\sigma + j\omega$ represents a complex frequency where the imaginary component $j\omega$ defines the frequency of oscillation and the real component σ defines the rate of decay (or growth) of that oscillation. Points along the $j\omega$ axis having a zero real component will therefore define points of sustained oscillation with neither decay nor growth. From this we can say that the system closed loop roots as viewed from the points along the $j\omega$ axis represent a measure of the frequency response of that system.

Although this concept was presented briefly in the Laplace transforms section, it bears repeating here using the above simple example. In conventional linear systems analysis, we typically obtain the frequency response of a system by substituting $s = j\omega$ into the closed loop transfer function. In our simple example, we obtain:

$$\frac{\text{output}}{\text{input}}(j\omega) = \frac{1}{(1 + j\omega T)}.$$

where the time constant $T = 1/K$. We can see clearly now that the process of substituting $s = j\omega$ is simply stating that we will constrain our view of the system response to the perspective of points on the $j\omega$ axis. Thus the general function $F(s)$ for any point in the 's' plane becomes a more specific function $F(j\omega)$ defining the system response to pure frequency.

Going back to our example consider now selecting various points along the positive $j\omega$ axis. We can say from the previous text on Laplace transforms that for this specific case with a single root the length of the vector from any point on the $j\omega$ axis to the root on the real axis represents both the magnitude and the phase response of the system to frequency excitation. When the selected point is at the origin of the plane representing zero frequency the vector becomes a reference vector corresponding to the steady state response of the system.

As we select different points (p) along the $j\omega$ axis as shown on Figure 4.30, the angle and length of the vector changes and the ratio between each of these vectors and the zero frequency (steady state) vector defines the frequency response of the system.

Now, since we are working with roots of the characteristic equation $(1 + loop = 0)$ and the characteristic equation is always in the denominator of the response equation, the frequency response is the reciprocal of the response vector divided by the reference vector. Thus the amplitude ratio is the length of the reference vector divided by the response vector

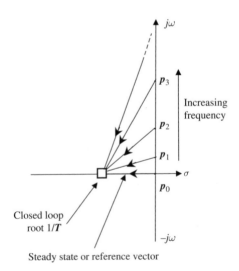

Figure 4.30 Vector ratio approach to frequency response

lengths at the various points along the $j\omega$ axis and the phase angle is minus the angle between the reference vector and the response vectors. Using this vector ratio method for determining the frequency response of this simple example we can make these observations:

- The amplitude ratio will be unity at zero frequency ($j\omega = 0$).
- As we move along the $j\omega$ axis to points p_1, p_2, etc. the amplitude ratio will attenuate towards zero at infinite frequency.
- The phase angle will be negative implying a lag which tends towards a 90 degree lag at infinite frequency.
- For the point on the $j\omega$ axis equal to $1/T$ radians per second, the phase lag is exactly 45 degrees and the amplitude ratio is $1/\sqrt{2} = 0.707 = -3\,\text{dB}$.

These observations are exactly in line with our conventional frequency response analysis of a first-order lag. From the 's' plane representation it becomes clear that as the closed loop root location becomes more negative, the time constant $1/T$ becomes smaller and the break frequency where the phase lag is 45 degrees now occurs at a larger value of $j\omega$ implying a faster responding system.

Let us now consider what happens when we add an additional lag to open loop response. This will add a second pole on the real axis and according to root locus construction Rule 2 there will be a locus on the real axis between the two poles and this locus will split into two separate loci with asymptotes at $\pm\,90$ degrees as shown in Figure 4.31.

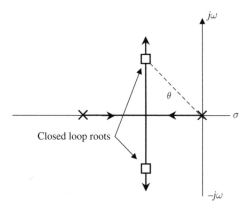

Figure 4.31   Root locus plot for a two pole open loop system

We now have a pair of complex conjugate roots in the closed loop system that can exhibit oscillatory behavior if the loop gain is increased sufficiently. However, in theory this system can never be unstable because the damping ratio $(\cos\theta)$ tends to zero but can only reach there at infinite frequency. Adding a third pole to the open loop system results in additional asymptotes for the loci because there are now three poles that need a zero as a target. In the first-order system with one pole the single zero target was located at minus infinity along the negative real axis. The two pole system required two zero targets and these are located at 90 degrees to the real axis at $\pm$ infinity. The three pole system requires three zero targets and these are located at minus infinity along the real axis and at angles of ±60 degrees to the real axis as shown in Figure 4.32 and in accordance with Rule 3 of root locus construction.

The trend is clear. The more lags in the open loop transfer function, the more poles reside in the left half of the 's' plane and the more aggressively the root loci move into the unstable right half plane. The above examples send an important message. The simple systems with a small number of open loop poles appear to be easy to stabilize. We must

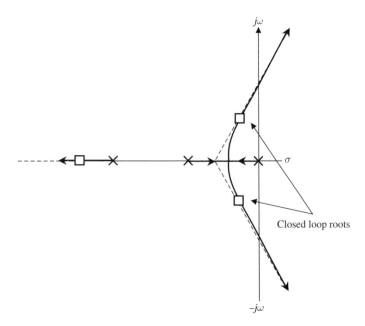

Figure 4.32 Three pole system root locus

remember that we often neglect the roots that are outside the frequency range of interest and, while their residues may be small, they still have a small but negative effect on the root loci by forcing them more quickly to the right and towards the unstable region. This phenomenon is similar to the observations made earlier about stiffness. It is always lower than expected because we never take all of the contributions into account and the result is lower than expected resonant frequencies. By the same token, we never take into account all of the time lag terms and therefore our root loci tend to impart optimistic results. This is why it is so important to design closed loop systems to have good stability margins and for the control system designer to be wary of system analyses that may have neglected or minimized the contributions of some of the faster roots.

Let us now consider the affect of adding zeros to the open loop transfer function. Figure 4.33 shows the three pole system with added zeros. The system on the left has a single zero on the real axis and the system on the right has a complex conjugate pair of zeros added. In each case the number of poles minus the number of zeros $(p - z)$, and hence the number of asymptotes, is reduced because the loci now have added targets within the plane. For the system with one added zero, the extreme left pole now produces a locus along the real axis to the zero and the asymptotes from the breakaway between the remaining two poles are now ± 90 degrees instead of ± 60 degrees. The added zero has therefore

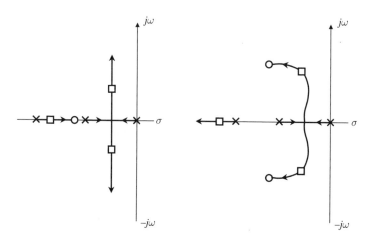

Figure 4.33 Three pole system with added zeros

pulled the primary locus to the left thus improving the damping ratio associated with the two complex roots.

The complex conjugate zeros on the right-hand system provide targets within the plane for the locus that breaks away from the real axis leaving just one asymptote that emanates from the farthest left pole to minus infinity along the negative real axis. Again a substantial improvement in damping of the primary closed loop roots is evident.

This review has provided the reader with an additional insight into the connection between the fundamentals of linear systems analysis developed earlier with the location of the open loop poles and zeros in the 's' plane and how this relates to the closed loop roots as the loop gain is increased. The section that follows will further reinforce the principles of root locus analysis as it is applied to the design of linear closed loop control systems.

4.6 Root Locus Example

Let us revisit the aircraft attitude control system analyzed in Section 2.7. This system was defined (using the 'D' operator) by the open loop transfer function (OLTF) which contains an integrator and three first-order lags:

$$\text{OLTF} = \frac{2.0}{D\,(1+0.02D)\,(1+0.1D)\,(1+2.0D)}.$$

We can now define the characteristic equation for this system based on the definition $(1 + loop) = 0$ using the Laplace transform notation:

$$1 + \frac{2.0}{s\,(1+0.02s)\,(1+0.1s)\,(1+2.0s)} = 0.$$

We now rearrange the equation so that the denominator terms are in the familiar $(s+p)$ form, i.e.

$$1 + \frac{2.0}{s\,(0.02)\,(s+50)\,(0.1)\,(s+10)\,(2.0)\,(s+0.5)} = 0,$$

i.e.

$$1 + \frac{500}{s\,(s+50)\,(s+10)\,(s+0.5)} = 0.$$

This system, therefore, has no zeros and four poles at 0, −50, −10 and −0.5. The gain term is not the actual loop gain but the loop gain divided by the product of each of the time constants associated with the poles. It is, however, directly proportional to the loop gain and may be referred to, loosely, as the loop gain in root locus analyses.

If we try to plot the poles in the 's' plane, it becomes clear quickly that three of the poles are reasonably close to the origin and the fourth pole is situated a long way away at −50. From the earlier discussions about residues at poles we said that if a pole is remotely situated from the main group its residue will be small as will its contribution to the response of the system. To illustrate this point Figure 4.34 shows the root locus plots for this system with and without the pole at −50.

The locus on the left has four starting poles on the real axis and since there are no zeros, there are two breakaway points for the four asymptotes that radiate from a point at −15.25 at +45 degrees and +135 degrees.

The alternative plot on the right is for the same system but with the pole at −50 neglected. Here there are only three poles on the real axis and only three asymptotes radiating from a point −3.5 at ±60 degrees

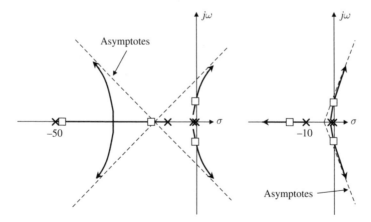

Figure 4.34   Root loci showing the effect of neglecting the remote pole

and −180 degrees. The rectangles shown on each locus are closed loop roots corresponding to the same loop gain indicating that there is no significant difference between the two approaches to the root locus analysis even though the root locus curves for the two approaches appear to be significantly different. We will therefore ignore the pole at −50 in the following discussions regarding this system.

Consider now that we need to have a control system that is both faster and with reasonably good damping. From the 's' plane diagram it would seem appropriate to introduce a numerator term with two complex conjugate zeros located at a higher value of $j\omega$ and with a more negative real component as indicated in Figure 4.35.

The new characteristic equation for this modified system is:

$$1 + \frac{k\left(s^2 + 5.4s + 30.33\right)}{s\left(s+10\right)\left(s+0.5\right)} = 0.$$

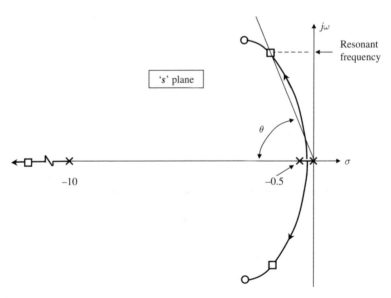

Figure 4.35 Attitude control system with second-order compensation

Factorizing the numerator gives us the complex conjugate roots which are the location of the zeros in the 's' plane, i.e.

$$1 + \frac{k\,[s+(5.4+4.8j)]\,[s+(5.4-4.8j)]}{s\,(s+10)\,(s+0.5)} = 0.$$

Now there are three poles and two zeros and the root loci will move from the two complex poles to the complex zeros and the third pole will move to a zero at minus infinity along the real axis as shown in the figure.

The closed loop roots are now vastly improved with the natural frequency of oscillation (indicated by the closest proximity of the complex roots to the $j\omega$ axis) of about 4 radians per second which is about four times the original value. The damping ratio of the complex roots, defined as $\cos\theta$ in the figure is about 0.4 compared with only 0.2 for the uncompensated system.

4.7 Chapter Summary

In this chapter we have developed a new visibility into the behavior of linear feedback control systems through the application of the Laplace transforms to define both the input forcing functions as well as the control system element transfer functions. This new visibility is afforded by the complex frequency domain which allows the control system analyst to understand the influence on the system dynamic behavior that the roots of the various elements in the open loop transfer function can impart due to their location in the complex frequency domain together with their location relative to the other elements in the loop.

We have also seen that we can readily replace the D operator with the Laplace operator 's' while retaining all of the application simplification benefits established in the early chapters of this book. From this point on, therefore, we will adopt the use of the Laplace operator 's' exclusively in all of our transfer function definitions. This is in line with the control engineering community at large where transfer functions are almost always expressed in Laplace terminology.

Chapter 4 also introduced 'root locus theory' as a design methodology that is particularly well suited to linear systems design analysis since it provides a means to observe the movement of a control system's closed loop roots in the complex frequency domain ('s' plane) as the loop gain is varied over the full range of values from zero to infinity. The insight provided by this methodology into the dynamic performance characteristics of linear closed loop control systems is powerful, and the practicing control engineer is encouraged to use this process as an analytical technique. It is not necessary to develop a rigorous analysis of the loci but to be able to draw the root locus from the system open loop transfer function quickly so that the closed loop root possibilities (and the impossibilities) can be visualized.

5

Dealing with Nonlinearities

This chapter addresses non-linearities and how to take account of them in closed loop systems analysis. So far we have assumed that the world is linear and continuous but unfortunately, more often than not, this is not the case. We therefore need to develop methods to accommodate nonlinearities as an extension of the linear analysis methods already established. These techniques use the concept of linearization where a system with nonlinear characteristics can be adequately represented by an equivalent linear representation. This approach does require some engineering judgment and we must always remember the guidance offered by Albert Einstein when he said that we should simplify the problem as much as possible . . . but no more!

Also addressed in this chapter are simulation techniques that allow the engineer to develop a computer representation of real world systems including their nonlinear characteristics and to evaluate the systems' dynamic behavior by observing the simulated responses to input stimuli.

5.1 Definition of Nonlinearity Types

Nonlinearities come in two fundamentally different forms. The first we can designate as continuous functions associated with mathematical expressions such as square roots and multiplication or division of two

Stability and Control of Aircraft Systems: Introduction to Classical Feedback Control   R. Langton
© 2006 John Wiley & Sons, Ltd

independent variables as well as the relationship between an input and an output transfer function that varies with the operating condition according to a predetermined curve or control law.

The second form of nonlinearity is the discontinuous form caused by such typical phenomena as static friction, deadband, saturation and hysteresis (also referred to in mechanical systems as backlash). Bang–bang control systems alluded to earlier also come into this category. Each of these nonlinearity categories can be treated separately with regard to stability analysis techniques. In the analysis of these discontinuous types of nonlinearity, the issue that we must recognize is that the system response will be dependent not only on the frequency of the stimulus but also on the amplitude of the input to the nonlinearities. Therefore we now have to consider two independent input variables when we evaluate nonlinear systems, first the frequency-dependent aspects of the system and second the amplitude-dependent aspects of the system.

Figure 5.1 shows an example of both of the two basic forms of nonlinearity described above. The figure shows the position versus flow characteristic of a typical hydraulic spool valve used to control the flow of hydraulic fluid into and out of a piston actuator. The graph in the figure is typical of a spool valve where the controlling spool land is slightly larger than the ports cut into the sleeve that it rides in. This is called 'deadband' and is favored by the hydraulic system designer as a means to minimize internal leakage of fluid from the pressure side to the return

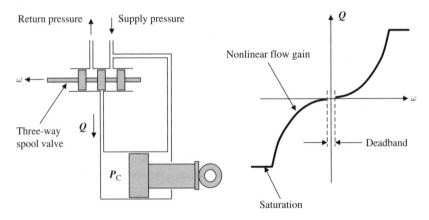

Figure 5.1 Typical spool valve characteristic

side with the attendant unwanted heat generation. This deadband, however, can be a problem to the control system designer because the flow gain of the valve is zero within the range of the deadband. The curve in the flow line is caused by the circular ports in the valve sleeve that produce a nonlinear flow area as the valve is moved away from the null position.

This nonlinearity can be avoided by cutting rectangular ports in the sleeve which is a commonly used but more expensive process. The third nonlinearity shown on this graph is the saturation feature that represents the effect of having a flow limiter in the hydraulic supply line or a travel limit designed into the error linkage. The above example serves to demonstrate that both forms of nonlinearity can (and often do) occur together and that this must be borne in mind when analyzing complex systems. The two forms of nonlinearity will now be addressed separately using common real world examples as a basis for discussion and explanation.

There is one additional form of nonlinearity that must be mentioned here and that is the pure time delay also referred to as a transport delay or dead time. This type of nonlinearity falls into neither of the above two categories and will therefore be discussed as a separate topic.

5.2 Continuous Nonlinearities

A good example of this form of nonlinearity is found in hydraulics and pneumatics where the flow equations involve a number of nonlinear functions. If we consider a simple flapper valve found in typical hydraulic servos (see Figure 5.2) we have a nonlinear relationship between the flapper displacement, and the flow through the nozzle.

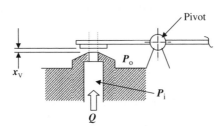

Figure 5.2 Flapper valve example of a nonlinear function

This type of equation does not readily fit into the linear transfer function methodology presented so far in this book, i.e.

$$Q = K_V x_V \sqrt{(P_i - P_o)}$$

where Q is the flow through the flapper valve, K_V is a valve flow/geometry coefficient, x_V is the flapper valve displacement, and $(P_i - P_o)$ is the valve pressure drop.

We have ignored here the flapper travel limits for the purpose of this exercise since this would fall into the category of discontinuous nonlinearities to be discussed in the next section. A block diagram of this equation would look like Figure 5.3. In order to reduce this problem to a linear equivalent where we can apply the analytical methods already described we use the method of 'small perturbations'.

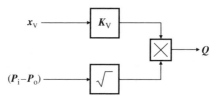

Figure 5.3 Flapper valve nonlinear block diagram

Here we consider what happens to the output if we disturb one input variable at a time while holding all the others constant. This is valid for small deviations around the operating point being considered. What we are doing in this example is to establish the partial derivatives of Q with respect to K_V and x_V and treating their contributions to the flow through the flapper valve Q independently. We can now rewrite the flapper valve flow equation using the prefix Δ to denote a small perturbation of the variable, i.e.

$$\Delta Q = \frac{\partial Q}{\partial x_V} \Delta x_V + \frac{\partial Q}{\partial (P_i)} \Delta (P_i)$$

(assuming that P_o is constant). We can now establish the derivatives:

$$\frac{\partial Q}{\partial x_V} = K_V \sqrt{(P_i - P_o)_0} \quad \text{and} \quad \frac{\partial Q}{\partial (P_i)} = \frac{K_V x_{V0}}{2\sqrt{(P_i - P_o)_0}}.$$

The $_0$ suffixes associated with the pressure and displacement variables denote the nominal values of each variable at the chosen operating condition. We can now construct a linear equivalent block diagram for the flapper valve (see Figure 5.4) that is valid for small perturbations about the selected operating point and usable by our linear analysis methods. This linearized block diagram implies that small variations in the flow through the flapper nozzle are equal to the sum of the effects of small perturbations in flapper displacement x_V and the pressure drop across the valve $(P_i - P_o)$ each considered separately.

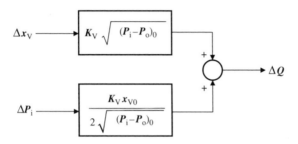

Figure 5.4 Flapper valve linearized block diagram

The continuous type of nonlinearity described above is also common in aircraft control systems where the operating conditions vary from sea level at low speeds to very high altitudes traveling at high speeds. These changes have a marked effect on the dynamics of the aircraft and its power plant. For the aircraft, the gains for the autopilot established for the sea level, low speed condition will be substantially different from the cruise condition if we want to maintain the same dynamic performance standards and stability margins. In the case of an aircraft gas turbine engine, the variations are not only introduced by the changes in flight condition but also as a result of variations in throttle setting. The following section uses the aircraft gas turbine engine fuel control as an example of the application of continuous nonlinearities in feedback control system design.

5.2.1 Engine Fuel Control System Example

A typical aircraft gas turbine fuel control system provides an excellent vehicle to demonstrate of the use of a continuous nonlinear function because of the natural benefits provided by this approach. The challenge here is to provide a means of compensation for both environmental variations (i.e. altitude and Mach number) and throttle settings that will result in a simple fuel control concept. To illustrate this problem, Figure 5.5 shows the significant differences in the sensitivity to changes in fuel flow that can typically occur between idle power and maximum power settings at a single operating condition (in this case at sea level static conditions).

In addition to the fuel flow sensitivity, the engine responds much more slowly at low power settings than at high power. An engine with a 1.0 second effective time constant at full power may have an idle power time constant of closer to 3.0 seconds. Add to this the effect of varying altitude and speed and we have a very complex process with the prospect of an equally complex control system. Therefore to control engine thrust as a measure of speed, for example, we need to accommodate the substantial variations in both the engine response terms (sensitivities and dynamics) as well as in the environmental conditions (altitude and speed) when developing the control system design.

A commonly used approach to solving this engine speed control system problem is to use compressor discharge pressure (P_C) as a feedback signal to the fuel metering unit to compensate for the variations in power setting by multiplying the speed governor output by this parameter. This makes a lot of sense because of the following two features.

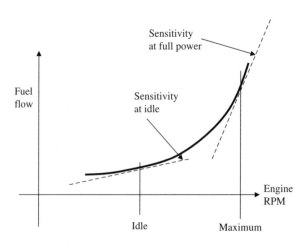

Figure 5.5 Gas turbine fuel sensitivity variation with throttle setting

- P_C feedback provides automatic correction to the fuel control system for changes in throttle setting *and* changes in flight condition (i.e. engine inlet pressure and temperature).
- It turns out that the ratio of fuel flow (W_F) and P_C is a fairly accurate measure of fuel/air ratio over most of the engine operating regime. Thus the ratio W_F/P_C remains fairly consistent throughout the operating envelope of the engine.

This control concept is illustrated by the block diagram of Figure 5.6. Here the P_C feedback signal to the fuel control is multiplied by the output from the speed governor to determine the required fuel flow to the engine. The governor gain K_G therefore, must be expressed in W_F/P_C ratio units per RPM of speed error.

The nice feature of this system is that the P_C multiplier effectively reduces the governor gain at low throttle settings when P_C is low and the engine sensitivity to fuel flow is high while increasing the gain at high power settings when P_C is high and fuel flow sensitivity is low. So what is the problem with this apparently ideal solution to the engine fuel control design problem? We will see this more clearly when we use the small perturbation method to linearize the system. When we linearize a multiplication function we hold one of the inputs constant and vary the other to establish the partial derivatives of each input leg. By inspection we can see that if the speed is held constant and we increase P_C we will increase fuel flow. This means that the P_C feedback is *positive* and not negative as is the norm in feedback control. Positive feedback is

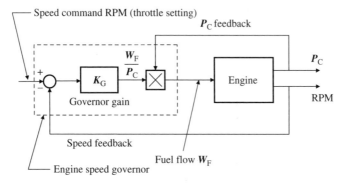

Figure 5.6 Gas turbine speed governor with P_C gain compensation

inherently destabilizing and therefore we need to look closely into the effect of P_C feedback with regard to the stability of the speed governor control loop.

To further evaluate this problem let us insert some numerical values into a block diagram that represent a gas turbine speed governor control loop at a specific power setting that is valid for small changes in accordance with our small perturbation methodology. This is shown in Figure 5.7 and is a very simplistic representation of a gas turbine engine but it does include the important characteristics that contribute to the speed control problem using a nonlinear P_C multiplier as a gain compensator.

The units used in this example are in common use in industry today, i.e.

fuel flow: lb/h
pressure: $lb/in.^2$ absolute
engine speed: RPM
W_F/P_C ratios: $lb/h/(lb/in.^2) = in.^2/h$

The compressor discharge pressure dynamics contains a fast path response to fuel flow changes and a slow path. The former is simply the immediate response to adding fluid into the combustion chamber and the latter is the change in P_C that occurs after the engine spools up (or down) in response to the change in fuel flow.

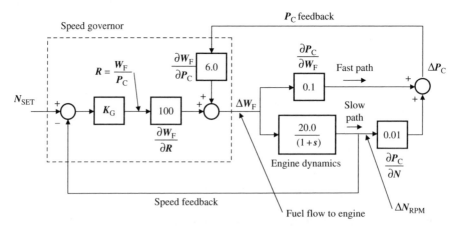

Figure 5.7 Gas turbine speed governor numerical example

From Figure 5.7 we can generate the expression for ΔP_C:

$$\Delta P_C = \Delta W_F(0.1) + \Delta W_F \left[\frac{20.0}{(1+s)} \right] (0.01).$$

This defines ΔP_C as sum of the fast path and slow path contributions. From the above expression we can define the transfer function relating ΔW_F and ΔP_C:

$$\frac{\Delta P_C}{\Delta W_F} = \frac{0.3\,(1+0.33s)}{(1+s)}.$$

We can now redraw the block diagram (see Figure 5.8) showing the P_C feedback as a separate inner control loop comprising the above transfer function in series with partial derivative $\frac{\partial W_F}{\partial P_C}$ which, in this example is 6.0 in.²h. This positive feedback loop can now be reduced to a single transfer function using the rule: forward path/(1 − loop), which yields:

$$\frac{\Delta W_F}{\Delta W_{FG}} = \frac{2.5\,(s+1)}{(s-0.5)}.$$

This shows that there is a positive real root in the closed loop transfer function indicating that this loop on its own would be unstable. We have

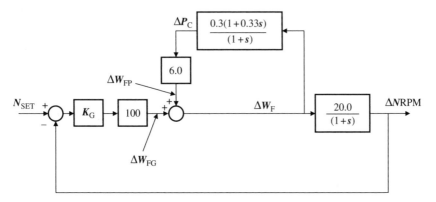

Figure 5.8 Rationalized speed governor and engine block diagram

an outer speed governor loop, however, that is able to provide an overall stable governor speed control as we shall see from the analysis that follows. To understand this let us develop the characteristic equation for the overall governor control system including both the inner and outer loops so that we can observe how the closed loop roots of the speed governor move as the gain is increased (i.e. let's look at the root locus plot for this system). The characteristic equation for this system is:

$$1 + K_G(100) \times (\text{inner loop}) \times \frac{20}{(s+1)}$$

i.e.

$$1 + \frac{K_G(100)(2.5)(20)(s+1)}{(s-0.5)(s+1)} = 0.$$

The 1 second time constant terms in the numerator and denominator cancel leaving a single root in the 's' plane at +0.5. The locus therefore goes along the real axis to a zero at minus infinity. Thus the system will be stable as long as the governor gain K_G is sufficiently large to ensure that the closed loop root is in the negative half plane.

Since we have only a single pole, the value of loop gain required for marginal stability is equal to the length of the vector from the pole to the origin, i.e. 0.5. This is the minimum loop gain for stability. If we wanted a governor with a break frequency of say 10 radians per second, this corresponds to a closed loop root at -10. The vector length from the pole to this point on the σ axis (and hence the loop gain for this solution) is 10.5. This defines $K_G = 0.0042$ ratios per RPM. As already mentioned this is a very simplistic interpretation of the engine control problem. In reality there will be additional lags around the loop which will cause the governor gain to be bounded as indicated in Figure 5.9 which shows the simple example locus and a more realistic form that would result from additional lags around the loop.

We have learned in this section that the way to handle nonlinear functions such as square roots, multiplication, division, variable gains, etc. is to use small perturbations about a specific operating point and to establish how the output from each nonlinearity would respond. With more than one variable in a nonlinear expression we consider the effects of varying each variable at a time while holding the others constant, i.e.

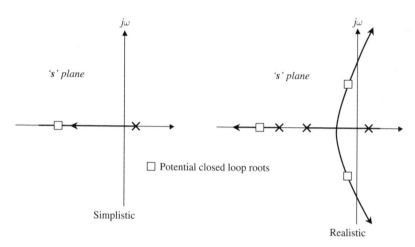

's' plane

's' plane

☐ Potential closed loop roots

Simplistic

Realistic

Figure 5.9   Gas turbine speed governor root locus plots

find the partial derivative of the output for each input variable in turn. The output is then the sum of the individual contributions. This gives us a linear system that we can analyze using the standard linear analysis techniques already developed.

It must be emphasized here that this technique is only valid for small excursions about a specific operating point and will not be representative of the dynamic behavior of the system in response to large disturbances. For this problem we use computer simulation which will be discussed later in this chapter.

5.3 Discontinuous Nonlinearities

Discontinuous nonlinearities differ from the previous type exhibiting sudden changes in response to input stimuli because of their inherent nature. Included in this category are deadband, saturation and hysteresis. These nonlinear features can be the result of any number of physical characteristics as illustrated by Table 5.1 which shows the characteristics and possible causes of some of the most common forms of discontinuous nonlinearity.

From the table there is one important observation that should be made regarding the nonlinearities described and that is with regard to hysteresis. This phenomenon is perhaps the most common function that the control systems designer must address and certainly the most

Table 5.1   Discontinuous nonlinearity characteristics

Nonlinearity	Example causes	Input vs output	Comments
Deadband	• Hydraulic valve overlap • Motor breakout torque		• Effective gain is reduced for small inputs • No phase shift
Saturation	• Valve flow limits • Travel limit stops • Amplifier voltage or current limits		• Effective gain is reduced for large inputs • No phase shift
Hysteresis	• Coulomb friction • Backlash		• Effective gain is reduced for small inputs • A phase lag is introduced that is larger for smaller inputs

troublesome. This is because of the phase shift that accompanies this nonlinearity. Both the deadband and saturation functions simply change the effective gain by modifying the signal transmission for either small or large signals, respectively, and therefore the accommodation of these nonlinearities is relatively straightforward. The effects of hysteresis, on the other hand, require careful study to ensure that the impact on system stability and dynamic performance are understood and acceptable.

In order to explain the behavior of the above nonlinearities we need to develop the output waveform that results from a sinusoidal input so that we can observe the signal transmission through the function. Figure 5.10 illustrates the saturation nonlinearity showing how a sinusoidal input is distorted as it is transmitted through the saturation element. From this we can see that the output waveform, while distorted due to the saturation effect, is in phase with the input signal. The effective gain (or transmission) through the nonlinearity is reduced and the bigger the input the bigger the loss (or gain reduction) of this nonlinear element.

One of the established methods for analyzing these effects involves the '*describing function*' which can be used to identify potential instability

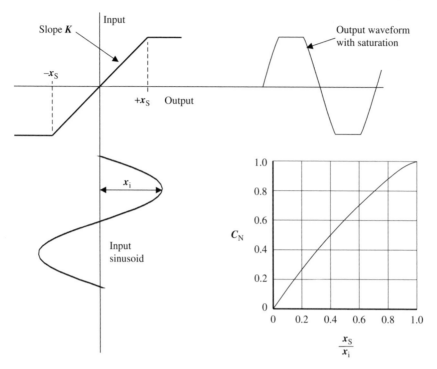

Figure 5.10 Saturation nonlinearity characteristics

for control systems with discontinuous nonlinearities based on sinusoidal inputs. This is convenient for the control systems engineer since frequency response is a well used and established technique for closed loop dynamic performance analysis. The describing function assumes that the outputs from nonlinear elements such as saturation, deadband, etc. can be represented by an equivalent sinusoid of the fundamental frequency of the input excitation. This is illustrated by the graph on Figure 5.10 which shows the waveforms and the describing function for the saturation element. In this case the describing function is simply a gain coefficient denoted by C_N that varies with the magnitude of the saturation limit X_S and the input amplitude X_i.

Since the transmission through the nonlinear element is not shifted in phase, C_N is a scalar quantity coefficient and therefore is very easy to apply to system stability analysis as we shall see later. We will not get into the geometric derivation of C_N here since we are more interested in the application of the describing function concept in control system performance analysis. For completeness, however, the equation for the saturation gain coefficient is shown:

$$C_N = \left[\frac{2\alpha}{\pi} + \frac{\sin(2\alpha)}{\pi} \right]$$

where $\alpha = \sin^{-1}\left(\dfrac{X_S}{X_i} \right)$.

Figure 5.11 shows the characteristics of the deadband nonlinearity including the describing function which, again, is simply a gain coefficient C_N.

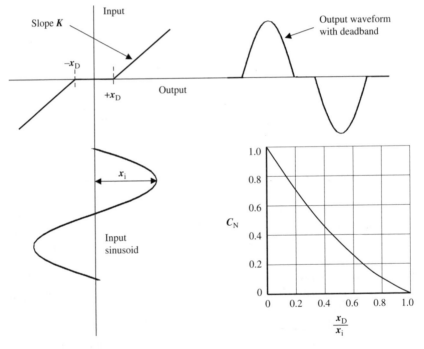

Figure 5.11   Deadband nonlinearity characteristics

Here the equation for the deadband gain coefficient is:

$$C_N = 1 - \frac{2\alpha}{\pi} - \frac{\sin(2\alpha)}{\pi}$$

where $\alpha = \sin^{-1}\left(\dfrac{x_D}{X_i} \right)$.

As with the saturation element the deadband nonlinearity has no phase shift associated with it and the effective gain is reduced as the input amplitude is reduced. The hysteresis element is more complex than the last two examples because it modifies both the gain and the phase shift in transmitting a sinusoidal input. As with all of the nonlinear elements described so far the effects on the control loop are dependent upon the amplitude of the input to the element and independent of the frequency of the input. We must take care to remember, however, that within a closed loop system the presence of simple nonlinearities such as the saturation and deadband elements described here can secondarily cause additional phase lag due to the fact that the loop gain has been changed by the presence of the nonlinearity within the loop.

Figure 5.12 shows the characteristics of the hysteresis nonlinearity which clearly indicates the phase shift of the output waveform relative to the ideal transmission waveform. The describing function for this element is no longer a simple gain coefficient but a complex number with gain and phase attributes. The graph shown on Figure 5.12 shows how

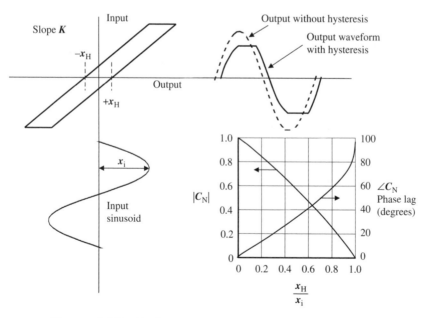

Figure 5.12 Hysteresis nonlinearity characteristics

the effective gain term $|C_N|$ and its associated phase lag $\angle C_N$ vary with the ratio of hysteresis half-width and input signal amplitude. It should be obvious to the reader that the potential for the hysteresis element to impact system stability is substantially greater than the saturation or deadband elements.

5.3.1 Stability Analysis with Discontinuous Nonlinearities

To understand the impact of discontinuous nonlinearities on closed loop system stability we need to go back to the condition for stability that we established in Chapter 2 where we defined the closed loop transfer function as:

$$\text{CLTF} = \frac{\text{forward\_path}}{(1 + \text{loop})}.$$

We also defined the condition for marginal stability from the characteristic equation, namely:

$$1 + \text{loop} = 0.$$

If we consider a generic closed loop system containing both linear and nonlinear elements as shown in Figure 5.13 where C_N the nonlinear element is dependent upon amplitude and independent of frequency and both $F(s)$ and $G(s)$ are linear dynamic transfer functions that

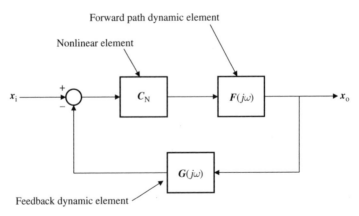

Figure 5.13 Feedback control system with nonlinear element

are independent of amplitude, we can write the closed loop transfer function:

$$\text{CLTF} = \frac{C_N F(s)}{[1 + C_N F(s)G(s)]}.$$

We must now substitute $s = j\omega$ since the describing function is valid only for sinusoidal response analysis. The characteristic equation is now:

$$1 + C_N F(j\omega)G(j\omega) = 0$$

or

$$F(j\omega)\,G(j\omega) = \frac{-1}{C_N}.$$

The term $F(j\omega)\,G(j\omega)$ is simply the open loop frequency response of the system which we are already familiar with. We can now plot this characteristic against the term $-1/C_N$ and see if the functions intersect.

The best graphical medium for demonstrating the stability aspects of control systems with nonlinear elements is the Nichols chart. This graph allows the depiction of the nonlinearity describing function as a single locus which incorporates both the gain and phase contributions of the nonlinear element. The importance of this will become apparent later as we discuss the hysteresis element. To begin let us consider the saturation element. The Nichols chart of Figure 5.14 shows how typical class 0, class 1 and class 2 control systems (containing zero, one and two integrators respectively in the open loop transfer function) relate to the locus of $-1/C_N$ which is a straight line along the open loop phase line of −180 degrees.

For the closed loop system to be unstable as a result of this nonlinear element, the frequency response locus must cross the $-1/C_N$ locus. It is clear from the chart that class 0 and class 1 systems will be unaffected stabilitywise by the saturation element because they cannot cross the open loop −180 degree line in the upper portion of the chart. On the other hand a class 2 system containing two integrators in the open loop frequency response starts with a −180 degree open loop phase lag and additional lags around the loop eventually force the response curve to cross the locus in order to bypass the instability point in a stable

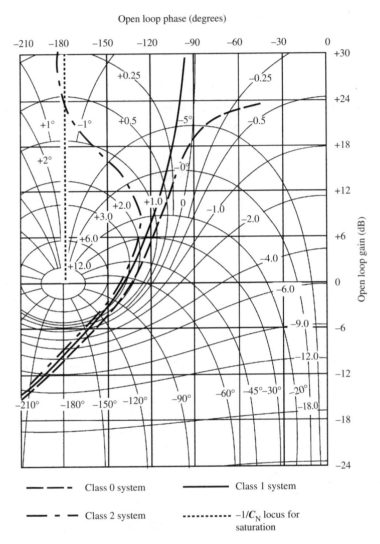

Figure 5.14 Nichols chart showing the effects of saturation

manner. At the intersection a stable limit cycle will occur at an amplitude defined by the ratio of the input oscillation to the saturation limit at the intersection. The same observation can be made regarding the deadband nonlinear element because the $-1/C_N$ locus is also a straight line along the −180 degree open loop phase line but with increasing amplitude going in the opposite direction.

When we look at the describing functions for both the saturation and deadband nonlinearities is becomes clear very quickly that:

- the only difference between the two loci are the points along the line that define the ratio of the input amplitude to the geometry of the nonlinearity;
- for most control systems saturation and deadband nonlinearities do not play a significant role *as serial elements on their own* in determining the stability margins of a typical control loop.

We need to be aware, however, that these apparently benign nonlinearities can turn into hysteresis once a loop is closed around them. One example of this is with regard to valve overlap (deadband) in a hydraulic servo. If the servo actuator has a friction load, this deadband can manifest itself as hysteresis between the input and the output. Thus an apparently benign nonlinearity is transformed into a more significant nonlinear element that can have a serious impact on the overall system performance and stability.

Let us now examine the hysteresis element and find out what makes this feature so troublesome to the control system designer. The Nichols chart of Figure 5.15 shows the locus of $-1/C_N$ for the hysteresis element together with the three typical frequency response plots for class 0, 1 and 2 linear control systems as before. Here, however, the nonlinearity locus moves towards the −90 degree open loop lag line as the input amplitudes to the hysteresis reduces towards the half hysteresis width, i.e. as X_H/X_i tends to unity.

We now have a situation where the class 1 and class 2 systems both cross the nonlinearity locus implying that a limit cycle will occur. Since the class 1 system starts its open loop phase at −90 degrees an intersection of the two loci is inevitable at some point. Therefore to reduce the impact of hysteresis on closed loop stability it is important to minimize the hysteresis as much as possible so that any limit cycling will be small. Also by ensuring that any additional lags around the loop have a high bandwidth, the frequency of any limit cycle can be kept very low to a point that it may be unnoticeable during normal control system operation.

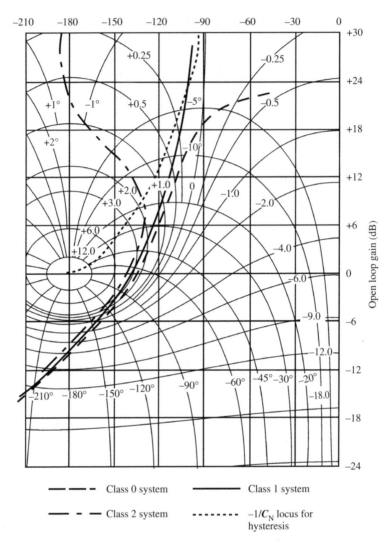

Figure 5.15 Nichols chart showing the effects of hysteresis

5.4 The Transport Delay

The transport delay is special type of nonlinearity where there is specific time delay before an event or control action is measured. A good example of this type of nonlinearity is the temperature control example shown in Chapter 3 where the control system mixed hot and cold fuels to achieve a desired temperature. Since the temperature sensor is located some physical distance from the mixing valve there will be a finite time delay before the mixing event and the measurement event. In this case the time delay will be equal to the distance from the mixing valve divided by the velocity of the fuel flowing down the pipe to the temperature sensor.

From a frequency response perspective, the transport delay adds a constant time delay to all frequencies therefore the phase lag will be directly proportional to the frequency of the input signal, i.e. doubling the frequency doubles the phase lag. Also the pure transport delay provides no attenuation as frequency increases. These two attributes make the transport delay something to be avoided wherever possible.

To demonstrate the powerful effect of transport delay, let us revisit the temperature control system analyzed in Chapter 3 beginning with the compensated solution which had an open loop transfer function:

$$\text{OLTF} = \frac{10}{s[1 + 0.05s(1 + 0.02s)]}.$$

This expression ignores the effects of any transport delay due to the location of the temperature sensor downstream of the mixing valve. For the operating condition analyzed, the transport delay can be calculated as the distance from the mixing valve to the sensor divided by the fluid velocity in the piping. For this example the transport delay was calculated to be 0.4 seconds.

Considering a sinusoid at 1 radian/second the time per cycle is 2π seconds. Therefore a shift of 0.4 seconds represents:

$$\frac{0.4}{2\pi}(360) = 23°$$

of phase lag. At 2 radians per second this phase lag doubles to 46° and so on.

Figure 5.16 Transport delay frequency response block diagram

The transfer function for the transport delay is e^{-sT_D} where T_D is the delay in seconds. To understand this, consider a sinusoidal input to a transport delay as indicated in Figure 5.16 where we have set $s = j\omega$. From this figure we can define the output from the delay which is:

$$f(t) = xe^{j\omega t}e^{-j\omega T_D} = xe^{j\omega(t-T_D)}.$$

This is clearly the input sinusoid shifted in time by the delay T_D.

Let us now develop the frequency response for the system including the transport delay of 0.4 seconds. The open loop frequency response transfer function is:

$$\frac{10e^{-0.4j\omega}}{j\omega\left[1+0.05j\omega\left(1+0.02j\omega\right)\right]}.$$

The Nichols chart of Figure 5.17 shows the original (compensated) response curve together with the response including the transport delay. This system is now unstable due to the substantial additional lag. Stable performance can be achieved by reducing the loop gain by about a factor of five (+14 dB). This would result in a gain margin of 6 dB.

This simple example demonstrates very clearly that the transport delay can have a substantial impact on closed loop system stability when it is large enough to add large phase lags in the frequency range where the open loop system characteristic crosses the 0 dB line. What makes the transport particularly difficult to deal with is that it brings no attenuation with it.

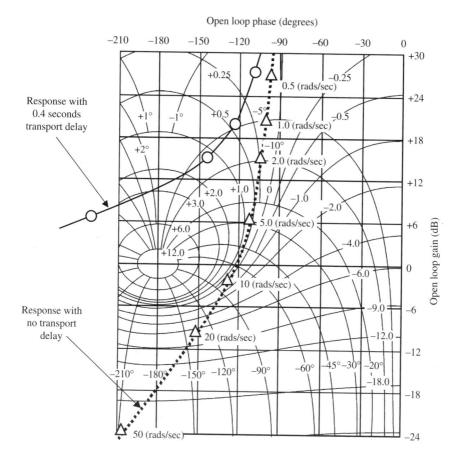

Figure 5.17 Effect of transport delay on temperature control system stability

5.5 Simulation

Simulation and modeling techniques are used by control system designers routinely as a means to verify or evaluate the dynamic behavior of control systems as an integral part of the design, development and certification process. This process provides a low cost, low risk means to test the system design before it is commissioned in the actual application. It also allows early identification and fixing of functional shortcomings, software bugs, etc. so that the system is, ideally, mature as it enters into service.

The aircraft simulator is an obvious example of the use of control system modeling that allows design engineers to optimize the aircraft dynamic characteristics and for pilots to critique the aircraft design and develop control skills long before the aircraft itself flies. In an aircraft simulator a computer is programmed to simulate the dynamics of the aircraft so that control inputs from the pilot generate a realistic simulated aircraft response which is fed back to the cockpit instrumentation to provide representative displays of attitude, altitude, speed, etc. In sophisticated simulators the cockpit is mounted upon a multi-axis platform so that pitch, roll and yaw motion can be created to give the pilot a more realistic ride. In this section we are going to confine ourselves to the use of simulation as a design tool for evaluating the dynamic behavior of control systems with nonlinearities that are otherwise difficult to analyze and evaluate.

Early simulation tools included analog computers where operational amplifiers were used to perform the basic summation and integration functions that make up differential equations. A typical analog computer of the 1960s would comprise up to 100 or more operational amplifiers with multiple inputs that could be configured as summing or integrating units. Potentiometers allowed the user to set specific system parameters and to 'scale' the equations so that 10 V represents, say, a specific velocity in m/s. A patch panel attached to the front of the machine allowed the user to connect up the various elements to represent the problem being evaluated. Figure 5.18 shows how our spring–mass system example would be set up on an analog computer. The potentiometers are adjusted to represent the values of mass, spring stiffness and viscous damping specific to the problem as well as the initial conditions for each integrator. Once operating, the voltage outputs from the two integrators will respond in a manner representative of the velocity and position of the mass. To see how the system responds to a step change in the input we would simply inject a voltage step at x_i and record the resulting response at x_o.

Nonlinear elements such as function generators and multipliers were also developed as part of the analog computers' capability to allow the user to easily investigate the effects of both continuous and discontinuous nonlinearities. In the 1970s these computers were further refined to include digital logic devices and in some cases a digital computer that could operate on one or more of the digitized analog voltages in order to solve complex equations 'on the fly' in parallel with the analog part of the machine. These devices were called 'hybrid computers'. One example where the digital section of the computer could be used to

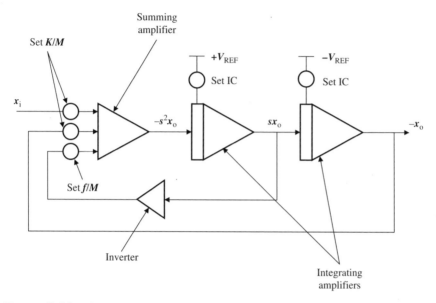

Figure 5.18 Analog computer model of the spring–mass system (IC stands for initial condition)

great benefit would be in computing pneumatic fluid flows where the transition between choked and unchoked flow is complex and highly nonlinear.

Today the analog and hybrid computers have been totally replaced by general purpose simulation tools that will run on both workstations and the faster of the standard PCs. These tools typically have libraries of elements, including all kinds of nonlinearities that can be integrated together in a user-friendly environment to make up a complete system. These tools solve the differential equations defined by all of the various elements by simply computing all of the variable derivatives for small time steps and then obtaining the output variables via a numerical integration routine. To demonstrate how simple this concept is let us develop the digital simulation code for the spring–mass system example.

First we must define the initial conditions for each independent variable, i.e. the velocity and position of the mass. We also need to define the values of the system parameters, i.e. mass M, the spring rate K, and viscous damping constant f. The basic computer code required to generate the time response of the mass position X_o (X0) resulting from a step change in the input X_i (XI) (using arbitrary values for the system parameters) would be typically:

```
C    INPUT=XI, OUTPUT=XO, OUTPUT VELOCITY=DXO
C    OUTPUT ACCELERATION=DDXO, PROBLEM TIME
C    STEP=DT
C    INPUT DATA
     M=100.0, K=400.0, f=80.0, T=0.0, DT=0.001
C    INITIAL CONDITIONS
     T=0.0, XI=0.0, XO=0.0, DXO=0.0
10   T=T+DT
     XI=1.0
     DDXO=((XI-XO)*K-f*DXO)/M
     DXO=DXO+DDXO*DT
     XO=XO+DXO*DT
     GO TO 10
```

This simple example uses rectangular integration to predict the value of the velocity (DXO) and position (XO) of the mass by assuming that the acceleration (DDXO) and velocity (DXO) respectively will continue at the same value for the time interval DT. Providing that the DT interval is small relative to the dynamic characteristics of the problem being modeled this is a reasonable assumption. If we plot out the graph of XO versus XI for the above computer model of our spring–mass system for every 100 time steps we would obtain the graph of Figure 5.19 which shows the response of a second-order system with a damping ratio of 0.2. This coincides well with our linear analysis methods.

An important point to note here is the choice of time increment (DT) used in the simulation. The value of DT must be sufficiently small relative to the bandwidth of the problem being simulated as to have no impact on the result obtained. A good rule of thumb in selecting a value for DT in digital simulation is to make sure that halving the value of DT does not change the response obtained.

The flexibility of digital computers and software make it easy to add nonlinearities into the model using conditional logic statements (if, then, else) and nonlinear functions such as multiplication, division, square roots and in fact anything that a digital computer can calculate. Going back to our spring–mass example let us replace the viscous damper with coulomb friction as illustrated by the block diagram of Figure 5.20.

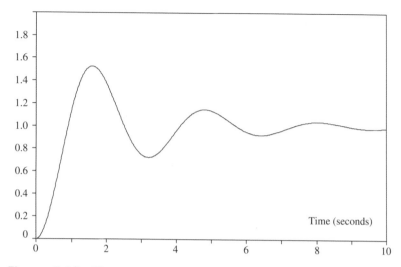

Figure 5.19 Step response of simulated spring-mass system

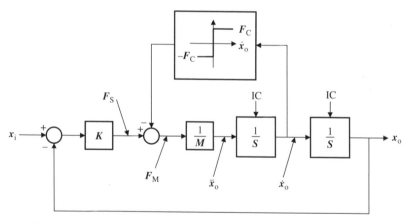

Figure 5.20 Spring-mass system with coulomb friction

This diagram applies only to the conditions where the velocity of the mass is not zero.

In our program logic, we need to be careful to ensure that when the mass is not moving the coulomb friction force does not act on the mass causing acceleration. The simulation must therefore do separate calculations for the moving state and the static state. The program code below shows how this nonlinearity can be represented including the condition when the mass is stationary.

```
C    INPUT=XI, OUTPUT=XO, OUTPUT VELOCITY=DXO
C    OUTPUT ACCELERATION=DDXO, PROBLEM TIME
C    STEP=DT
C    INPUT DATA
     M=100.0, K=400.0, FC=70.0, T=0.0, DT=0.001
C    INITIAL CONDITIONS
     T=0.0, XI=0.0, XO=0.0, DXO=0.0, FS=0.0, FM=0.0
10   T=T+DT
     XI=1.0
     FS=(XI-XO)*K
     IF(ABS DXO<0.0001) THEN
C    FM FOR ZERO VELOCITY CONDITION:
     IF(FS<FC) THEN FM=0.0
     DXO=0.0
     ELSE FM=(ABS FS-FC)*SIGN FS
C    FM FOR NON-ZERO VELOCITY CONDITION:
     FM=FS-FC*SIGN DXO
     DDXO=FM/M
     DXO=DXO+DDXO*DT
     XO=XO+DXO*DT
     GO TO 10
```

Figure 5.21 compares the step response of the linear system with the nonlinear version with coulomb friction. As shown in this figure, coulomb friction introduces quite a different response as a result of the amplitude dependency of the friction term. Note also that the coulomb friction version ends up with a steady state error.

The problem for most practicing engineers is that generating bug-free software code to represent accurately complex nonlinearities and other custom features of a control system can be a challenge for the nonprogrammer. This point is demonstrated by the relatively complicated code necessary to define such a common nonlinearity as coulomb friction. The good news is that the software tools available today take that issue away by providing proven modules (libraries) for such elements as coulomb friction, hysteresis, saturation, nonlinear functions, etc., that can be selected and installed into a control system model by the control system analyst in a user friendly manner.

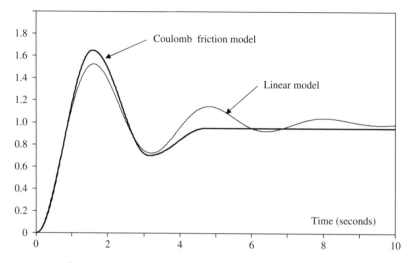

Figure 5.21 Comparison of step responses

We can use the linearized engine governor analysis presented earlier in Section 5.2.1 as an example of the value of simulation in control system design and analysis. In the earlier section the method of small perturbations was used to evaluate the stability of the speed governor. This analysis was only valid for the selected conditions and did not provide any visibility as to the behavior of the system for large excursions in speed demand. Simulation allows the designer to assemble dynamic models that are fully representative of the physical system including multiple types of nonlinearities and over a wide range of operating conditions.

The block diagram of Figure 5.22 shows a model of a gas turbine engine that can be used to examine control system behavior over the full power range. The fuel flow input W_F is compared with the steady state fuel flow W_{Fss} derived from the engine steady running line function to obtain the amount of over (or under fueling) ΔW_F. This is multiplied by the derivative $\Delta Q/\Delta W_F$ (which varies with rotational speed N) to obtain the excess torque to accelerate the engine spool. Compressor discharge pressure P_C is generated in a similar manner. The 'fast path' change ΔP_C resulting from the over (or under) fueling ΔW_F is added to the steady state value P_{Css} to get the complete dynamic variable P_C.

This model approach is one of the simplest methods and depends on having good engine performance data available. More rigorous models

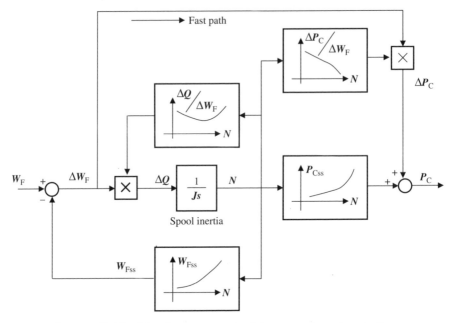

Figure 5.22   Nonlinear gas turbine engine simulation

based on the complete aero-thermodynamics of the compressor and turbine are also in common use throughout the aerospace industry. Also not shown in the model presented here is the application of corrected variable techniques that allow easy compensation for changes in engine operating conditions, i.e. altitude, temperature and Mach number.

The value of the full range nonlinear model allows the control system designer to evaluate the performance of the control system better not just from the point of view of the stability at specific power settings but also during the transient conditions between, say, idle and maximum power. This is important for gas turbines because the acceleration and deceleration process must be carefully controlled to prevent compressor surge during acceleration and flame out during deceleration.

Figure 5.23 shows the schematic of a fuel controller showing how the speed governor, acceleration and deceleration limiters are combined.

By combining the non-linear simulations of the engine and control system the full range dynamics of the system can be readily evaluated including the transition from acceleration (or deceleration) limiting to the speed governing mode. This is illustrated on Figure 5.24 which shows a typical slam acceleration plot of fuel flow and engine speed.

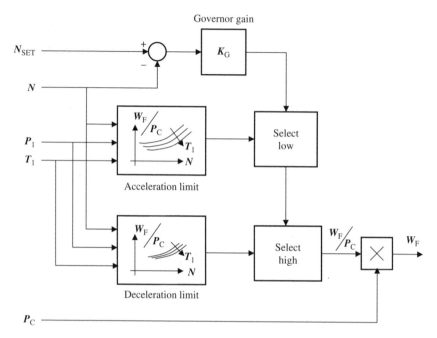

Figure 5.23 Fuel control nonlinear schematic

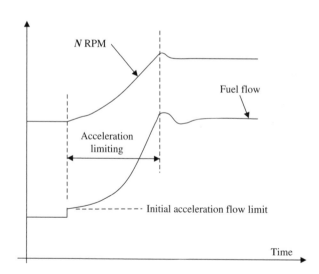

Figure 5.24 Typical slam acceleration plot

A slam acceleration is a step change in throttle setting usually from idle to maximum power. At the start of the plot, the engine is running on the speed governor and as the step change is applied there is a sudden increase in fuel flow which is limited to the W_F/P_C acceleration limit associated with the prevailing speed and inlet temperature and pressure conditions (N, P_1 and T_1); this is multiplied by P_C to give the fuel flow required by the engine. As the engine accelerates the acceleration limit continues to increase until, as the new set point is neared, the governor path output becomes lower than the acceleration limit and the speed governor takes over control.

This is a good example of the value of simulation as a tool to provide visibility into the dynamic behavior of highly nonlinear systems. Simulations such as this are used to validate actual control system hardware and software during development before using an actual engine. Thus control bugs can be identified and fixed safely and without jeopardizing expensive hardware.

5.6 Chapter Summary

Dealing with nonlinearities is a continual challenge to the control system engineer because the real world is, more often than not, nonlinear and the analytical tools that can be readily applied to closed cool system problems are based on the assumption of linearity. This chapter describes a number of techniques that can be used to work around this issue, the foremost being 'linearization'.

Nonlinear types are described as being in three categories:

- continuous nonlinearities;
- discontinuous nonlinearities;
- the transport delay.

Continuous nonlinearities such as square root functions, multiplication, division, variable gain functions, etc. can be treated as linear elements using the small perturbation technique that states that for small perturbations around a specific operating point, systems with such nonlinearities can be easily transposed into an equivalent linear form that is readily analyzable using the standard linear method. Discontinuous nonlinearities such as saturation, deadband and hysteresis, require a different approach called the 'describing function' that allows the amplitude dependent behavior and frequency dependent behavior of a system to be evaluated separately and the stability impact of the nonlinearity, if any, to be identified in both amplitude and frequency terms. This method of nonlinear analysis is somewhat limited in that it can only handle one nonlinear element at a time and the dynamic behavior in the presence of the nonlinearity is not easily visualized. Of the various types of discontinuous nonlinearity described, the hysteresis element was established as the most troublesome due to the phase lag effect on signal transmission that tends to 90 degrees as the input signal amplitudes approach the hysteresis width.

The transport delay was singled out as a special case that generates a phase lag that is directly proportional to frequency and zero attenuation. Dealing with this element is easy, however, since it can be included as a serial element in the open loop transfer function and its impact on stability can be readily assessed.

Finally, simulation and modeling as a technique for evaluating nonlinearities was described covering the history of analog, hybrid and digital simulation tools. The availability of general purpose software packages that are easy to use even on today's standard PCs were described as the standard used by industry. It is stressed, however, that it is always good practice to do an approximate linearized analysis before getting into complex models so that the analyst has a reasonable understanding of what to expect from the outcome of the simulation exercise.

6

Electronic Controls

Critical aircraft control functions such as flight controls and engine power management were, by necessity, controlled using systems with mechanical interconnections and mechanisms because of the criticality of the function itself. Loss of control for any reason would likely mean loss of the aircraft (when something goes wrong at 30 000 ft you can't just 'pull over').

Early attempts during the 1950s to use the obvious benefits afforded by the flexibility of electronics were fraught with reliability problems due to the inability of electronic components of the day (such as thermionic valves/vacuum tubes) to withstand the hostile environment (primarily vibration and temperature effects) associated with the aircraft flight environment. Following the early failure of electronics to meet reliability expectations in the field, innovations involving the use of hydromechanical and pneumomechanical technologies were developed to a high degree. This was particularly true with regard to engine control technology.

This temporary diversion brought with it some exceptional technologies including 'fluidics' which were envisioned at the time as the answer to the environmental reliability problems of electronic controls. With the advent of the transistor and its miniaturization leading to very large scale integration (VLSI) technology and eventually single chip microprocessors, a revolution in the involvement of electronics throughout the aerospace industry began that continues to this day.

Stability and Control of Aircraft Systems: Introduction to Classical Feedback Control   R. Langton
© 2006 John Wiley & Sons, Ltd

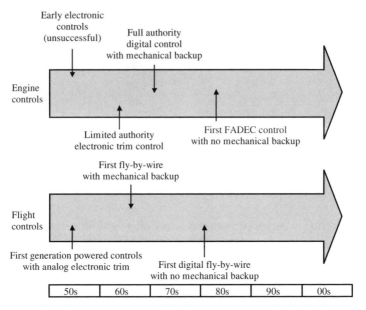

Figure 6.1 History of aircraft electronic controls technology

To give some perspective to this commentary, Figure 6.1 shows some
of the major technology milestones regarding the application of electronics in the control of aircraft engine and flight control systems.
By the 1980s digital electronic control was becoming the accepted technology standard for almost every control function in both military and
commercial aircraft programs of that decade. The popular acronyms of
the day were FBW (fly-by-wire) for flight control systems and FADEC
(full authority digital engine control) for engine control systems. In each
case the safety criticality issue was addressed by the use of multi-
channel redundancy and in some cases the use of a mechanical backup
system. Today every new aircraft or engine program accepts that digital
electronic control is the only viable approach to be considered due to
the affordability, reliability and computing power of modern electronic
hardware coupled with the functional benefits afforded by the flexibility
of software. There remains, however, an important challenge which is to
bring software reliability and system-level functional maturity at entry
into service for complex electronic control systems up to the expectations
of the investors and users of this powerful new technology.

Before we come to the application of digital computers in modern control technology, which is the primary goal of this chapter, the following section is included which describes the use of analog electronics in closed loop control systems which became extremely popular from the 1960s through the 1980s. While digital electronics has emerged as the technology of choice today, analog solutions are still in common use where the cost of software design, development and maintenance for aircraft applications may be considered prohibitive for relatively simple control functions.

6.1 Analog Electronic Controls

Analog electronic controls in aircraft applications were enabled largely by the advent of the transistor and solid state electronics technology. During the 1960s and 1970s before the availability of general purpose single chip microprocessors, the control system designer looked first to analog electronics technology to provide an easy way to implement custom dynamic functions. We saw the introduction of auto-stabilizers in flight control systems providing limited authority trim functions that added to the pilots mechanical commands. Engine controls saw the introduction of the 'supervisory control' as a means to reduce pilot workload and to enhance the maintenance intervals by preventing inadvertent exceedances such as transient over-temperature events that can seriously reduce the engine hot section life.

Two of the most ambitious accomplishments of analog control technology developed in the 1960s were the first aircraft Auto-land system developed by Smiths Industries and the quadruplex flight control system for the USAF F-16 aircraft developed and produced by the Astronics Division of Lear Seigler. These analog control systems were eventually replaced by digital electronic controls.

As the growth in electronic technology continued there emerged an increase in the application of hybrid electronics technology wherein analog electronics was mixed in with the emerging digital technology. This was a stepping stone towards the application of custom digital computer designs later developed by the large corporations who were prepared to absorb the expensive tooling costs associated with custom electronics in an attempt to obtain a competitive position and also to fill a technology vacuum that continued to grow. The section that follows summarizes the key functional features of analog electronic controls in order to give the reader a basic understanding of the principles involved in their design

6.1.1 The Operational Amplifier

The most significant driver in this area of electronics was the availability of the transistorized 'operational amplifier' in the early 1960s. The term 'operational amplifier' or 'op-amp' implies the use of high gain amplifier circuits to perform common mathematical operations such as summation, integration and differentiation. Operational amplifiers have key functional properties that allow the control system designer to 'build' dynamic transfer functions using passive devices such as resistors and capacitors. These important properties are:

- extremely high voltage gain typically greater than 10^5;
- operational bandwidth well into the megahertz range.

These characteristics allow the designer to neglect the dynamics associated with the amplifier safely, leaving the passive devices to determine the functionality of the control circuit.

Operational amplifier control functions are determined by the input and output impedance circuits around the amplifier as shown in the generic schematic of Figure 6.2. Schematics of analog amplifier control functions typically do not show the amplifier, power supply and ground

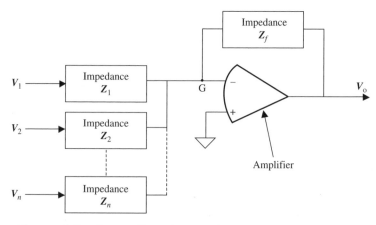

Figure 6.2 Operational amplifier generic schematic

circuitry which are usually taken for granted. The key aspect of the schematic of Figure 6.2 is the concept of the 'virtual ground' at the junction of the input and feedback impedances (point G on the figure). This is brought about by the very high amplifier gain which ensures that the input to the amplifier at point G must always be essentially at ground potential. Thus the output voltage relative to the virtual ground is simply the sum of the input voltages multiplied by the ratio of the feedback and input impedances, i.e.

$$V_{\mathrm{o}} = -\left[V_1\left(\frac{Z_f}{Z_1}\right) + V_2\left(\frac{Z_f}{Z_2}\right) + \cdots V_n\left(\frac{Z_f}{Z_n}\right) \right]$$

The negative sign reflects Kirchoff's Law that the current flowing *into* the virtual ground G must equal the current flowing *out* from G to the output.

6.1.2 Building Analog Control Algorithms

The control and compensation algorithms developed in Chapter 3 can be easily assembled using operational amplifiers and passive devices. The simple integrator is obtained by using a capacitor for the feedback impedance as shown in Figure 6.3.

From basic electricity theory, the current flow through a capacitor is defined by the equation:

$$V = \frac{1}{C}\int \mathrm{i}\,\mathrm{dt}$$

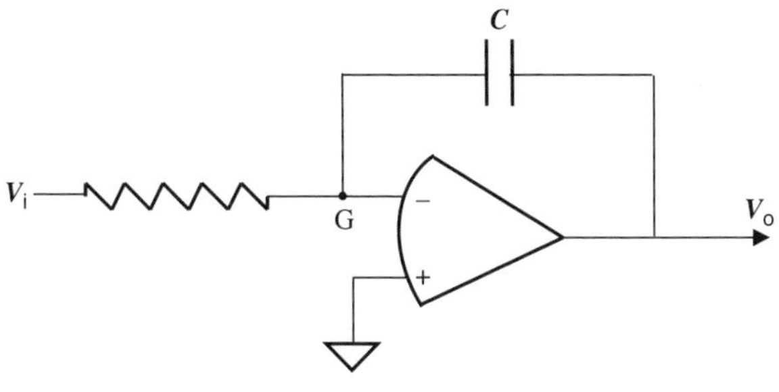

Figure 6.3 Integrator schematic

which in Laplace operator format is the same as:

$$V = \frac{1}{C}\left(\frac{i}{s}\right).$$

From this we can see that the impedance represented by a capacitor becomes:

$$Z = \frac{V}{i} = \frac{1}{Cs}.$$

We can now equate the current flow through the input resistor and the feedback capacitor to give us the relationship between the input voltage V_i and output voltage V_o as follows:

$$\frac{V_o}{V_i} = -\left(\frac{1}{RCs}\right).$$

This is the transfer function of an integrator with RC having units of time. Using this approach control and compensation can be easily applied to operational amplifier circuits. Figure 6.4 shows examples of this for proportional plus integral and lead–lag functions.

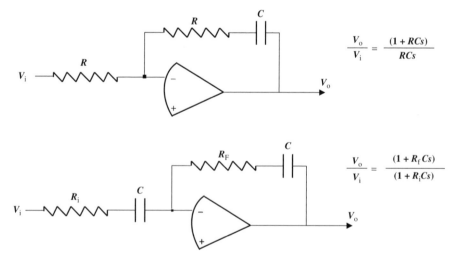

Figure 6.4 Proportional plus integral and lead–lag circuits

The above provides a brief illustration of how analog electronic circuits can be assembled to emulate the dynamic transfer functions used in the control and compensation of feedback control systems. Due to the extremely fast response of the underlying electronics associated with these operational amplifier-based circuits, there is no significant dynamic penalty involved in the control implementation. The digital controller, however, is different in that there is a finite, and often significant, time delay involved in providing the loop closure and control algorithms involved. The rest of this chapter is therefore focused on the digital electronic controller.

6.2 The Digital Computer as a Dynamic Control Element

So how does a digital computer operate as a dynamic controller within a closed loop control system? This issue may not be obvious to the non-specialist because the digital computer is associated with high-speed number crunching in response to program instructions that are typically not time dependent in the sense of providing dynamic response to external time-dependent processes.

The explanation of this question lies in the following three items which together make the digital computer viable as a control element:

- the real time clock which ties the computer to a real world measure of time;
- the analog-to-digital converter (A-D converter) which provides the ability to convert analog signals at high speed into the digital domain;
- the digital-to-analog converter (D-A converter) which transposes the output of the computer's calculations back into the analog world.

6.2.1 Signal Conversion

In order to be effective as a controller the digital computer must have high speed access to all forms of input signals including analog voltages, pulse trains, serial data streams and discretes. Similarly the computer has to communicate the control output signals back into the physical

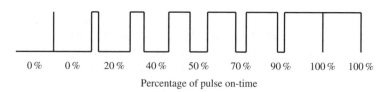

0%   0%   20%   40%   50%   70%   90%   100%   100%

Percentage of pulse on-time

Figure 6.5 Pulse width modulation (PWM) example

world so that the appropriate control action can be initiated. Output signals can also be in analog, discrete and serial data form. Pulse width modulation (PWM) is also a common form of output format used in digital systems using a carrier square wave and varying the width of each pulse as a measure of the magnitude of the output signal (see Figure 6.5). The percentage of pulse 'on-time' is referred to as the pulse width 'duty cycle' that can vary from 0 to 100%.

The A-D converter takes a number of forms the most common of which is the successive approximation type. This method is illustrated by the diagram of Figure 6.6 which shows the conversion of an analog voltage of 5.8 V (an arbitrary value) in a signal range of 0.0 to 10.0 V. The conversion takes place by using the mid-point of successively smaller ranges as a 'guess'. If the guess is low a 1 bit is set, if it is high a 0 bit is set. In the example shown the process is:

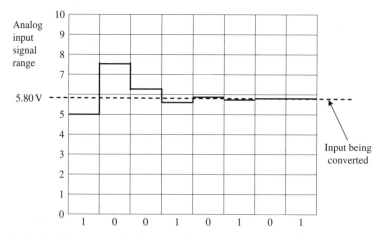

Figure 6.6 Successive approximation A-D conversion

(1) The first guess is half of the full range (i.e. 5 V). In this case the guess is low so the most significant bit set is a 1.

(2) The second guess is at the mid-point of the upper half of the signal range (i.e. 7.5 V). In this case the guess is high so a 0 bit is set for the next most significant bit.

(3) The third guess is the mid-point of the third quarter of the signal range (i.e. 6.25 V). In this case it is still high so a 0 is set.

This process continues for the number of significant bits of the A-D converter. In the example shown, the eight bit successive approximation process results in the binary number 10010101. This in turn is equivalent to:

$$(128 + 0 + 0 + 16 + 0 + 4 + 0 + 1)/256 = 149/256$$

For a 10.0 V signal range this is $(149/256)^*10 = 5.8203125$ V.

From a signal accuracy standpoint, an eight-bit A-D converter has a maximum potential error in the conversion process of 1 in 256 which is about $\pm 0.4\%$ of full scale. In most systems a converter with a 10 bit resolution or better is used which reduces the error to less than 0.1% which is good enough for most applications.

The D-A converter uses the same principle in reverse. The computer sets a series of switches representing the value of the binary number to be converted and a summing amplifier with weighted input resistors adds the contributions of each bit to achieve the analog equivalent output as indicated in Figure 6.7.

Pulse trains from magnetic pick-ups are typically used to sense shaft speeds. These sensors can be easily converted into the digital domain using counters and clocks. A pulse train generated from a magnetic pick-up mounted on a rotating shaft can be accumulated in a digital counter so that over a fixed time period the binary number in the counter represents the rotational speed of the shaft.

The problem with this interface is the fact that for a specific time window, the number of pulse counts in a fixed time period becomes less as the shaft speed increases. Also the critical operating condition is usually the high-speed condition. As the shaft speed reduces the pulses per conversion increases providing superior resolution at operating conditions where it is typically not required. One way around this problem is to use the incoming pulse train as the start and stop bits for a high-speed clock. This way the resolution throughout the speed range is much improved.

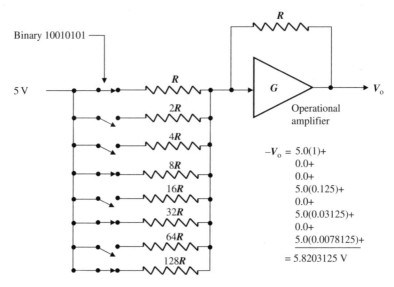

Figure 6.7 D-A converter example

There are many other innovative approaches to this problem that can be adopted which are beyond the scope of this book as an introduction to the subject of feedback control. However, the fundamentals presented here provide a good starting point from which the subject can be pursued further.

Serial data transmission and reception is an equally important aspect of digital control that must be covered here. The most common serial data transmission format in use in the aircraft industry today is the ARINC 429 data bus (ARINC stands for Air Radio Inc and is a USA standard for avionics). This standard has been in commercial service for more than 30 years and has even been adopted by the military community as a result of the push for dual-use technology in new military programs. This specific standard employs a unidirectional data bus that transmits data, labels and parity information using a 32 bit protocol. The ARINC 429 data stream is a 0.0 to $+5.0\,\text{V}$ square wave that provides a time series of 1s and 0s from one source to one or more receivers. Two standard transmission rates are available within the standard, one at 12.5 kilobits/second (kbps) and the higher speed version at 100 kbps. The label identifies the specific variable being transmitted and 18 bits of the 32 bit word are allocated as data thus providing adequate resolution for almost any application.

Serial data must be converted from a serial, time-based format into a parallel format that can be used by the digital controller. This activity is typically performed by dedicated hardware that communicates with the central processing unit (CPU) by setting a flag or a discrete bit that tells the CPU that a serial data word has been converted into an equivalent parallel format and ready for transfer. The CPU can then transfer this data to a predetermined memory location in the random access memory (RAM) for use by the control algorithm. When a conversion is in progress the conversion bit is not set. There are many other serial data standards available with different protocols and speeds. For most applications the specific standard(s) for serial data communication will be specified by the aircraft manufacturer.

6.2.2 Digital Controller Architectures

The issue that faces the control system designer is the fact that the computer requires a finite amount of time to do these conversions and to execute the control algorithms required by the system control. Figure 6.8 is a simplified schematic of a single channel digital controller having 'n' input variables with 'm' outputs. The single line arrows depict analog signals while the broad arrows represent a parallel bus

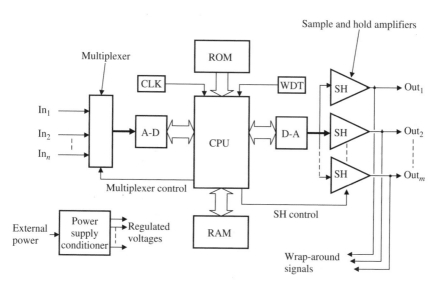

Figure 6.8 Simplified schematic of a digital controller

typically 16 bits (or 32 bits) wide. The multiplexer is a solid state switch that allows the CPU to select each input signal in turn to the A-D converter. After processing the input data in accordance with the instructions stored in the read only memory (ROM), the D-A converter is loaded with the required outputs. The RAM is used by the CPU as a scratch-pad for temporary storage of variables and intermediate calculations. This sequence of signal conversion and control logic computation continues with the output changing in response to the input changes via the control algorithm. The output amplifiers are 'sample and hold' devices that hold the last D-A output until it is updated at the end of the next iteration. To verify the correct operation of the D-A and output amplifiers, the output signals are wrapped around back to the input multiplexer so that the CPU can verify correct operation of the complete output stage. The clock (CLK) tells the CPU when to start a new computer iteration and the watch-dog timer (WDT) shuts the computer down if it is not reset at the end of each clock cycle. This ensures that undetected program errors or physical failures of the computer electronic hardware cannot result in indeterminate control actions.

In a typical digital control system the activities of the CPU can be broken down into two major functions:

- the input-output signal management (I/O handler);
- the control logic execution.

The I/O handler is responsible for the A-D and D-A processing and all of the continuous built-in-test (BIT). This includes range checks on all of the inputs as they are digitized, along with test and ground signal conversions that ensure correct and accurate functioning of the input interface. The I/O handler loads the appropriate values for each input into predetermined locations in RAM and loads the outputs from predefined addresses in RAM into the D-A converter.

The control logic is simply the execution of the control instructions stored in ROM. The values of the input variables are always located in the same address in RAM under the control of the I/O handler. This separation of I/O management and control logic execution minimizes the impact of changes during the system development and certification process. Within the computer logic is an executive control that manages the time sequencing of the CPU activities. Typically the tasks associated with one complete iteration are broken down into a number of small time segments as indicated by the diagram in Figure 6.9. For example,

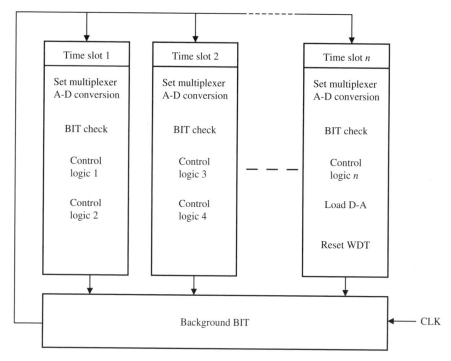

Figure 6.9 Typical control executive arrangement

a complete iteration time allocation may be 20 ms which may be broken down into say, 20 segments of 1ms. Within each 1ms time slot, a number of tasks are allocated including signal conversion, BIT checks, control logic, etc.. After completing each time slot task allocation, the CPU executes secondary BIT referred to as 'background checks' which are completed on a time available basis. In this category are RAM read-write tests and program memory check sum calculations to verify the integrity of the computer. The background activity continues until the real time clock initiates the next computation cycle.

A brief discussion of redundancy is appropriate at this point. Where the controlled functions are safety-critical as is typical with most of the major control systems in aircraft, the probability of loss of control must be acceptably small. The probability of a catastrophic event must be lower than one chance per 10^9 flight hours to satisfy the certification authorities and to achieve such levels of safety may mean providing more than one control channel so that in the event of a failure in flight a second channel with equivalent control capability can take over to maintain the safety of the control function.

Figure 6.10 shows a number of control architectures from single channel to a quadruplex (four) level of redundancy. Clearly the higher the level of redundancy the lower the probability of loss of the control function. This of course comes at a price which includes hardware cost, weight, reliability and maintainability. There is also an additional burden associated with additional redundancy and that is the software complexity associated with redundancy management and fault

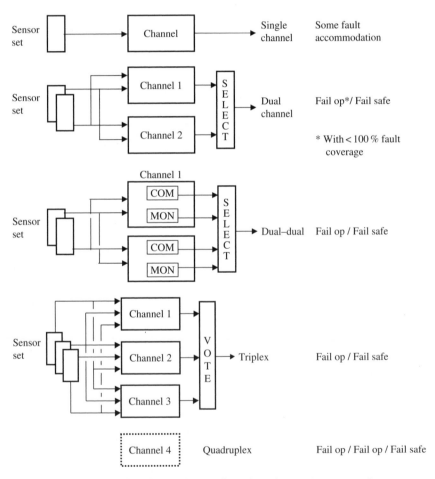

Figure 6.10 Overview of redundancy concepts

detection and accommodation. The simplex or single channel system does have the ability to evaluate the health of the system elements and, following the detection of a fault, accommodation control logic can be introduced that can allow continued operation with some degradation in performance. There are, however, a significant number of faults that can be catastrophic for the system the most obvious of which is loss of the computer itself where the watch-dog-timer could shut down the system.

The dual channel approach offers a much more functionally reliable solution than the simplex approach since we have doubled up on the sensors and computing hardware. The issue that now becomes critical is the level of fault coverage that can be attained with this architecture because an undetected failure can result in loss of control function. In a dual channel system levels of fault coverage significantly above 90 % are difficult to achieve and verification of this coverage factor can be subjective. One of the most challenging issues with dual channel systems is the detection of 'signal in-range faults'. While the majority of transducers and interface circuit faults result in an out-of-range condition that can be readily identified, the in-range fault results in a difference in value between the two channels. In this case either one of the signals can be correct or incorrect. The selection of dual channel architecture, therefore, can only be tolerated in systems that can safely tolerate loss of function, albeit with a very low probability of occurrence. Early FADEC solutions for aircraft engine control used dual channel architectures because the contribution of the FADEC to the engine shut-down rate was considered acceptably small at about two shut-downs per million flight hours.

As the miniaturization revolution in digital electronics continued with substantial improvements in the size, cost and reliability, the dual–dual architecture won favor over the traditional dual channel system. The addition of a second computer into each channel allowed close to 100 % fault coverage to be achieved. Here a command processor (COM) and a monitor processor (MON) execute identical software. Any differences between the COM and MON processor outputs will result in that channel being taken offline. The in-range sensor failure issue remains, however, even with this more sophisticated architecture. For systems where loss of function can be catastrophic such as in-flight control systems, triplex or even quadruplex architecture becomes necessary. Here any single failure can be voted out by the remaining two channels providing 100 % functionality after any single failure. The quadruplex architecture provides even more safety protection by accommodating two failures without loss of functionality. Even the in-range sensor failure can be identified via the voting process.

6.3 The Stability Impact of Digital Controls

We now need to examine how the digital controller can affect the dynamics of the control loop. Figure 6.11 shows how a digital controller with sample and hold output stages transmits a sign wave. In the example shown the output is refreshed about 12 times in one complete oscillation of the sine wave. Inserting specific values this is equivalent to a controller with an update rate of 20 ms transmitting a 4 Hz sine wave.

The staircase effect on the output shows an approximate sinusoidal output that is displaced in time. As the frequency of the sine wave input increases (i.e. fewer signal updates per oscillation) the time shift increases and the output waveform becomes less sinusoidal as shown in Figure 6.12. The effect of the signal digitization with the sample and hold output can be approximated by the following transfer function:

$$\frac{e^{-Ts}}{(1+Ts)}$$

where T is the computer update rate.

This is a transport delay of Ts in series with a first-order lag with a time constant also equal to the computer update rate T. This approximation can be easily verified. Going back to Figure 6.11 and drawing in a best-fit sine wave to the quantized output indicates a phase lag of about

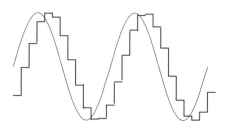

Figure 6.11 Transmission of a sine wave through a digital controller

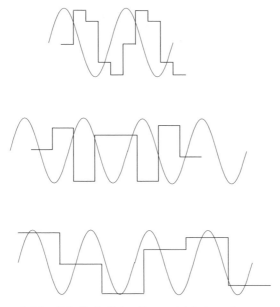

Figure 6.12 Digital distortion at higher frequencies

one sixth of a cycle or 60 degrees. Substituting $T = 0.02\,\text{s}$ and $\omega = 25.0$ radians per second (4 Hz) into the above transfer function the phase lag is calculated:

$$\text{transport delay phase lag} = \frac{(0.02)\,(25)}{2\pi}\,(360) = 28.7°$$

$$\text{first-order lag phase} = \tan^{-1}(25)\,(0.02) = 26.6°.$$

This gives a total calculated phase lag of 55.3 degrees which agrees closely with our graphical approximation.

The above example demonstrates that the dynamic characteristics of a digital controller can contribute significantly to the stability of closed loop systems by virtue of the phase lag resulting from the digitization and computing delays inherent in the process. In designing closed loop control systems with digital controllers, careful attention must be given to the signal acquisition and output update rates. One of the best examples that demonstrates many of the key issues involved in digital control is the FADEC (full authority digital engine control) for the gas turbine engine. Without getting into the redundancy issues and sticking to the basic control functions let us consider the engine control system arrangement shown in Figure 6.13 which is a simplified hardware diagram typical of a commercial turbofan engine fuel control system. The gearbox driven from the main engine shaft drives the high pressure fuel pump, the fuel metering unit and a dedicated alternator to provide power to the FADEC unit. The FADEC controls the fuel flow metered to the engine and, simultaneously the position of inlet guide vanes (IGV) on the compressor inlet. Figure 6.14 shows the signal flow around the two closed loop functions.

The FADEC receives throttle commands from the flight deck and autothrottle and the prevailing flight conditions are available via a serial data bus. Feedback signals from the engine include fan speed, N_1, gas generator speed N_2, compressor discharge pressure, P_C, turbine gas temperature, T_{GT}, and two position feedbacks, one from the fuel metering valve, X_{FV} and one from the IGV actuator output, X_{IGV}. As is typical with complex control systems the dynamic response requirements associated with the various control modes can differ substantially, for example the temperature and pressure signals respond very quickly to changes in fuel flow while the engine shaft speeds have a much slower response.

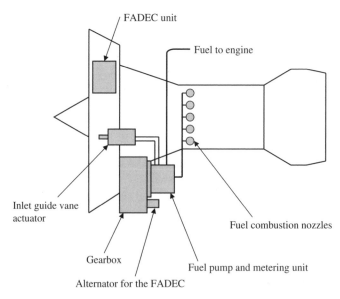

Figure 6.13 Engine control system hardware

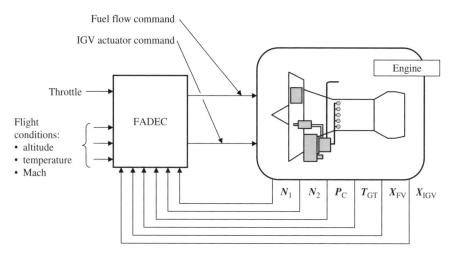

Figure 6.14 Engine control system signal flow diagram

The inlet guide vane (IGV) actuator must respond quickly in order to do its job of controlling the airflow into the compressor during large power transients.

So how does the control system designer cope with these varying dynamic response demands from the digital controller perspective? The easy answer is to require that the digital controller will support the highest bandwidth control loop and that all signals must be accessed and processed at that same high-speed rate. This 'brute force' approach is not only a very inefficient use of data and processing power, the throughput required may be significantly beyond the capacity of available (not to mention affordable) processors. A more effective solution is to develop a control executive that accesses data and processes the control logic at update rates that are compatible with stability and control needs but no faster.

In the engine control system example there are large differences in the signal rates of change among the various input and feedback variables. At the slowest extreme are the external inputs which provide throttle commands and flight operating condition data to the FADEC. These variables need not be accessed by the controller any faster than, say, every 0.1 s. The next slowest responding variables are the engine shaft speeds. These variables change relatively slowly due to the inertia of the rotating machinery. An update rate of 20 ms is adequate here. The engine pressures and temperatures are fast moving variables and should be accessed by the controller at, say, 10 ms intervals.

This leaves the position feedback signals from the fuel metering valve and the IGV actuator output plus the output commands to the engine for fuel flow and IGV position. These control loops should have band-widths of 10 Hz or better in order to be dynamically separated by at least a decade from the process they are controlling (in this case the engine). By separating the roots of the control system from those of the process we are minimizing any undesirable dynamic contribution from the controller during transient operations. This is related to the comment in Chapter 5 when we discussed the effect of the location of the open loop system roots on the closed loop system behavior. Here we said that poles located well away from the main group will have small residues.

Based on these observations therefore the feedback signals and the output commands should be refreshed at a faster rate than the other control parameters. Experience has demonstrated that an update rate of 5 ms in an application similar to the one described here is adequate to ensure good stability margins. A good rule of thumb is to select an update rate for a digital control loop such that its inverse in cycles per second is separated by at east a decade from the bandwidth of the loop being controlled. In this example the inverse of 5 ms is 200 cycles per second which is well separated from the control loop bandwidth of about 10 Hz. From this exercise we can see that we have bought a lot of spare computing capacity by managing the signal conversion and output update rates based on their specific dynamic contribution to, and involvement in, the overall system.

6.4 Digital Control Design Example

To reinforce the important features associated with digital closed loop system stability, this section goes through the process of establishing the design parameters of a simple closed loop system application with a digital controller. We will assume for the purpose of this example that the complete control loop can be represented by the linear transfer functions shown in Figure 6.15 together with a digital controller whose task is to close the loop and apply the appropriate control action to the process being controlled.

As shown, the process is represented by an integrator in series with two first-order lags and the output from the process is measured by a

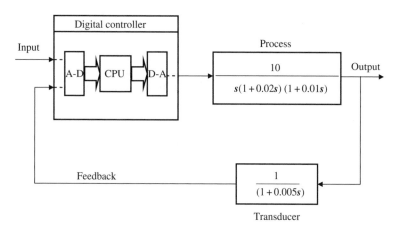

Figure 6.15 Digital control design example schematic

transducer with an additional 5 ms lag. The control system performance requirements are:

(a) to have a steady state error of less than 2% of the operating point for inputs in the range 10% to 100% of full scale;
(b) to have a bandwidth of 1 Hz or better measured as a closed loop phase lag of less than 50 degrees at 1 Hz;
(c) to have stability margins in excess of 40 degrees phase and 10 dB gain.

The steady state accuracy requirement allows us to determine the signal conversion resolution of the A-D converter. Even though we have an integrator in the loop which ensures that the steady state error must always be reduced to zero as a result of the integral action, we must now take into consideration the resolution of the digitization process.

For example, if we select an eight-bit A-D converter we must recognize that the conversion process has a resolution of no better than one part in $512(2^8)$. This means that the converted signal could have an error of up to $\pm 0.4\%$ of the input signal range. Therefore for an input signal of 10% of the signal range a 0.4% of full scale error equates to 4% of the operating point with twice the allowable error based on the accuracy requirement of (a) above. A 10-bit A-D converter has a resolution of one part in $2048(2^{10})$ which correlates to an error of $\pm 0.05\%$ of full scale. Therefore an input signal of 10% of the full range will have a conversion error of no more than $\pm 0.5\%$ of full scale thus providing a 2:1 margin over the accuracy requirement.

The output D-A resolution requirement is much less critical in this example because the process contains an integrator and therefore the D-A output is commanding a rate-of-change to the process. The resolution of this signal, therefore, does not contribute to the steady state accuracy requirement since it must always be zero in steady state. Consideration may be given by the control system designer to using only an eight-bit D-A device as a cost saving approach.

When we consider the control loop design the natural place to start is to evaluate the linear solution, i.e. ignore the digital dynamics in order to determine the controller function that best fits the requirements. Therefore let us develop the open loop response for the system as though it were a linear open loop system. The open loop transfer function assuming a simple gain term in the controller is:

$$\text{OLTF} = \frac{G\,(10.0)}{s\,(1+0.02s\,(1+0.01s))}$$

where G is the controller gain.

Figure 6.16 shows the open loop response on a Nichols chart for a value of G equal to 1.0 indicating good stability margins (16 dB gain margin and 65° of phase margin) a system bandwidth of about 10 radians per second or 1.5 Hz. Thus the system as a traditional linear system satisfies the dynamic requirements assuming that the digital controller has no impact on the loop dynamics.

To establish the digital controller cycle time requirement we can consider the crossover frequency which in this case is 10 radians per second and determine what the cycle time would be to reduce the phase margin to the specification limit of 40 degrees. i.e. we solve for T in the equation:

$$\angle\left(\frac{e^{-10jT}}{1+10jT}\right) = 25°.$$

If we calculate the phase angle for a cycle time of $T = 0.02\,\text{s}$ we get 22.9 degrees which suggests that with this signal update-rate the system will have a phase margin of slightly better than 45 degrees. The Nichols chart of Figure 6.17 shows the system response curve with the digital controller dynamics included showing that the dynamic response

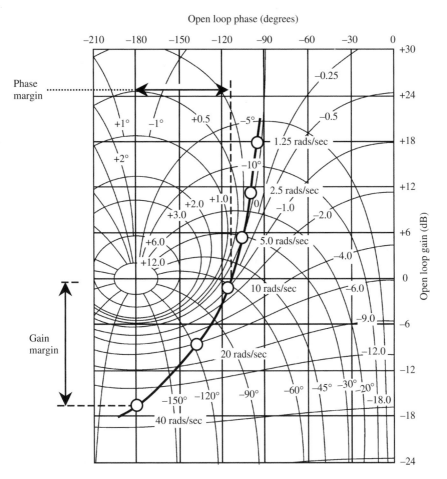

Figure 6.16 Nichols chart of the equivalent linear system

requirements established at the outset are met. If the available computer cannot support the update-rate required, then the loop gain must be reduced to achieve the specified stability margins. The impact of this would be to reduce the system bandwidth hence degrading the closed loop dynamic response.

This example is deliberately very simple to ensure that the fundamental issues are understood. In the real world a digital controller would not be selected to do a simple comparison and gain multiplication for a simple linear system. The power of the digital controller

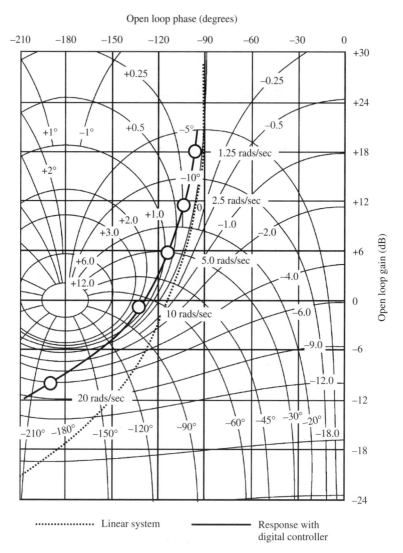

Figure 6.17 Impact of the digital controller dynamics

is its ability to apply very complex control algorithms which may be necessary to control a highly nonlinear process with widely varying gains and dynamic characteristics over the system operating envelope. The fundamental design principles are the same, however, even though there may be many different operating points to consider in a similar linearized fashion. The following section provides an introductory insight into how dynamic compensation algorithms can be programmed into the digital controller.

6.5 Creating Digital Control Algorithms

Having acquired the digitized signals defining process output requirements, process feedback and other signals related to the prevailing operating conditions, the digital controller's task is to compute the output drives for each control loop. This task typically involves the generation of dynamic elements that emulate linear transfer functions in order to provide the appropriate control action and/or signal compensation for optimum dynamic performance of the system. For example:

- integral control;
- proportional plus Integral control;
- proportional plus derivative control;
- lead–lag compensation, etc..

These dynamic elements and many others can be generated from three basic building blocks, namely, the integrator, the first-order lag and the derivative elements. The digital representations of these elements are developed below.

6.5.1 The Integrator

This function can be very easily achieved using rectangular integration as shown in Figure 6.18 by adding the last computed value of the

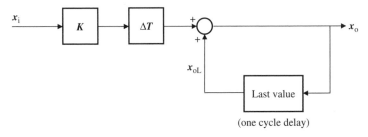

Figure 6.18 Simple integrator algorithm

output to the product of the newly computed derivative and the cycle time, i.e.

$$X_o = X_{ol} + X_i K \Delta T$$

where: X_o is the new output, X_{ol} is the output computed by the last cycle, X_i is the input to the integrator, K is the integrator gain in s^{-1} $\left(\text{second}^{-1}\right)$ and ΔT is the controller cycle time in seconds.

By inspection it can be seen that provided the derivative (input) term is finite the output will continue to grow and that the growth rate will be proportional to the magnitude of the derivative term.

6.5.2 The First-order Lag

To generate this function we simply add a feedback around the integrator algorithm as indicated in the block diagram of Figure 6.19. In this case the time constant of the lag $T = 1/K$. To demonstrate this concept in detail, Figure 6.19 also shows the computed results in tabular and

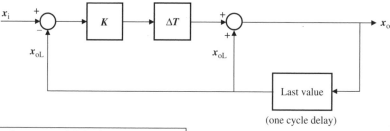

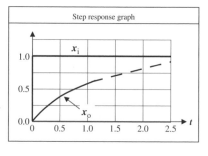

Step response tabulation for $K = 1.0$, $\Delta T = 0.1$

t	x_i	$(x_i - x_{oL})$	x_{oL}	x_o
0.0	0.0	0.0	0.0	0.0
0.1	1.0	1.0	0.0	0.1
0.2	1.0	0.9	0.1	0.19
0.3	1.0	0.81	0.19	0.271
0.4	1.0	0.729	0.271	0.344
0.5	1.0	0.656	0.344	0.370
0.6	1.0	0.630	0.370	0.410
0.7	1.0	0.590	0.410	0.469
0.8	1.0	0.531	0.469	0.522
0.9	1.0	0.478	0.522	0.570
1.0	1.0	0.387	0.570	0.613

Figure 6.19 First-order lag algorithm overview

graphical form for a first order lag for values of $K = 1.0$ and $\Delta T = 0.1$ following a step change in the input X_i from 0.0 to 1.0 at $T = 0.01\,\text{s}$. The results show a good representation of a first-order lag step response. The quantization process does introduce errors which are functions of the ratio between the time-constant being modeled and the cycle time of the controller. In the example shown the error is about $0.2\,\%$ (the output value at $T = 1.0\,\text{s}$ should be close to 0.63 to agree with linear theory (the algorithm calculated a value of 0.613).

As a general guideline the smallest time constant must be at least an order of magnitude larger than the controller cycle time. If this rule is violated very large errors can result and even instability of the algorithm itself.

6.5.3 The Pseudo Derivative

The preferred algorithm for this function is in fact a combination of the integrator and first-order lag elements. As indicated by the heading this is not a pure derivative element but a differentiator with high frequency filtering via a first-order lag. This approximation provides close to pure derivative action at frequencies below the break frequency of the first-order lag while providing an all-important high frequency filtering effect to guard against the generation of unwanted high frequency noise. As was mentioned earlier, differentiation is an inherently noisy process and therefore the use of pure derivative action should be avoided. The transfer function of this pseudo derivative element is:

$$\frac{Ts}{(1 + Ts)}.$$

This effect is generated digitally by putting an integrator with a gain of $K = 1/T$ in the feedback path around a simple gain of 1.0 as indicated in the diagram of Figure 6.20. As shown in the graph of the step response, the output responds immediately to the step change in the input and then decays to zero.

The digital control algorithms described here are just brief summary in order to give the reader an idea of what is involved in generating dynamic control elements within a digital controller. Clearly the possibilities are almost unlimited and as computer power continues to increase the ability to generate essentially instantaneous transmission through the digital controller will become close to reality.

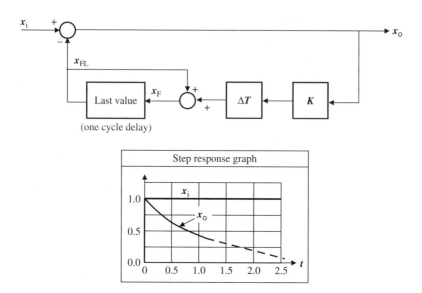

Figure 6.20   Pseudo derivative algorithm overview

6.6 Chapter Summary

The digital computer as the control element within a feedback control system is rapidly becoming the preferred solution as a result of the exponential growth in computing power together with an equally impressive reduction in cost. With the flexibility afforded by software, the control possibilities are almost infinite.

It behooves the control system engineer, therefore, to be aware of the issues involved in applying this powerful technology to the task of the system controller. This chapter provided a top-level explanation of how the digital computer fulfills the role of system controller with regard to signal and data acquisition, control logic computation and control output commands. A brief introduction into digital control architectures provided insight into the issues of fault detection and accommodation and the levels of redundancy necessary to ensure continued function following one or more failures, since this is the most important issue with regard to control system operation in an airborne situation.

The impact on closed loop stability as a result of the time delays associated with the computing function was explained and a simple way to represent digital computer dynamics via an equivalent linear element was described and reinforced using a simple control system example.

Finally, some examples of control algorithms that can be programmed to emulate the standard linear transfer functions were developed and some general guidelines were established in order to avoid common pitfalls associated with quantization and cycle time that will ensure that the algorithms behave in accordance with the intended linear definition.

7

Concluding Commentary

The objective of this chapter is to review the contents of the book and to summarize the most important aspects of the material presented making sure that the reader is well armed with the tools necessary to appreciate the subtleties of the design, analysis and testing of closed loop control systems. It should be recognized that this book is not intended to make control theory experts out of its readers but to create a heightened level of awareness of the consequences of feedback control and of the contribution that certain elements around the loop can make to the dynamic behavior of a system. The emphasis throughout, therefore, has been to provide the potential practitioner with a number of basic tools that rely heavily on graphical methods to provide a quick insight into the response characteristics of control systems.

One successful outcome from digesting the content of this book would be to allow the reader to delegate complex control systems analysis work to the control engineering specialist department while, at the same time, having an expectation of the likely outcome from this analysis so that a meaningful discussion of the results can occur. This ability has been referred to several times earlier as having a 'feel for the problem' and can be likened to using a calculator to obtain the answer to a complex calculation and having an expectation of approximately what the answer should be. All too often, today, users of sophisticated analytical tools blindly write down the answer without having any idea as to its validity. The recommendation is, therefore, to use the knowledge presented in the

Stability and Control of Aircraft Systems: Introduction to Classical Feedback Control   R. Langton
© 2006 John Wiley & Sons, Ltd

previous chapters to establish a basic understanding of the system using lots of block diagrams and sketches. This comment will be expanded upon in the following paragraphs.

7.1 An Overview of the Material

The first step in the knowledge gathering process presented in Chapter 1 was to lay the groundwork for some of the basic mathematical concepts that are needed in order to express the concept of response and oscillatory behavior. Here we introduced the concept of the D operator and the use of block diagrams to show graphically the interaction between the various elements around the control loop. A refresher on the subject of complex numbers provided the background necessary to appreciate the mathematics behind the oscillatory response of physical systems such as the spring–mass system that was used as an example.

The early chapters focused on the use of frequency response, which is perhaps the most commonly used analytical and test methodology used by the control engineering community, as a means of classifying the dynamic response of closed loop systems and with the expression of amplitude ratio in dB plotted against log frequency we learned that it is easy to establish the product of all the elements around the loop by simply adding the gain in dB for each element. Total phase can also be determined in a similar manner by simple addition. With practice the generation of frequency response graphs is quick and easy thus giving the practitioner an immediate insight into the stability margins of the system.

The next challenge was to simplify the task of moving from open loop response, which is used to establish stability margins, to the closed loop response in order to observe how the system responds to changes in the set point or from the application of external disturbances. To support this process we introduced the Nichols chart which should be considered as a key graphical aid to illustrate both the open loop and closed loop characteristics on a single diagram. Not only does this graphical tool provide immediate visibility into the closed loop response, it eliminates the mathematical tedium associated with the generation of closed loop roots.

The compensation techniques expounded in Chapter 3 provided the reader with some insight into the design process whereby dynamic elements can be introduced into the control loop to compensate for some of the undesirable dynamic features of the system, be they the process itself or one or more of the transducers used to measure the various states of the process.

Laplace transforms were introduced in Chapter 4 as a tool supporting the development of transient response solutions for linear systems using simple algebra together with the use of standard tables to represent the various forcing functions and output responses. This chapter included a significant amount of mathematics in order to describe the logic associated with the transformation process; however, the application of the principles of Laplace transformation can be integrated into the response analysis process without bringing with it the need to remember the mathematical origin of the process when solving specific response analysis problems. The most important aspect of the Laplace transform tutorial was the introduction of the s plane or 'complex frequency domain'. This unique feature allows the analytical engineer to see where the control system roots reside and what their dynamic contribution to the real world will be. This point is illustrated in Figure 7.1 which shows the complex frequency domain (i.e. the 's' plane) and how the locations of the roots of a typical second-order system affect the real world response.

When the roots are close to the $j\omega$ axis the real world response to inputs close to the undamped natural frequency are greatly magnified. As the real component of the root location becomes more negative,

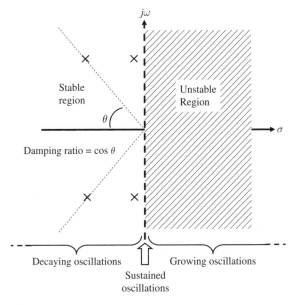

Figure 7.1   The complex frequency domain summarized

the oscillation tendency quickly disappears so that when the cosine of the angle θ is 0.707 or greater, there is no measurable oscillation. The region to the left of the $j\omega$ axis is the stable region for system roots with decaying oscillations while to the right side of the $j\omega$ axis is the unstable region where oscillations continue to grow. Roots lying exactly on the axis will exhibit sustained oscillations.

It is an interesting paradox that in the complex frequency domain, the real world is represented by the imaginary ($j\omega$) axis while the real axis determines the rate of decay (or growth) of oscillations based on the remoteness of the system roots from the $j\omega$ axis. The ability of the reader to become comfortable with this 's' plane concept is important in order to develop that all-important 'feel for the problem'. From this point we were able to replace the D operator in transfer functions with the Laplace operator s since this has essentially the same meaning while giving the additional insight provided by the s plane representation of the system. In fact it is typical for control engineers to consider the s and D operators as interchangeable and while this is not mathematically correct it is nevertheless common practice. Certainly when we set s (or D) equal to $j\omega$ for frequency response analysis this is, in fact, mathematically correct.

Chapter 5 showed how nonlinearities can be taken into account using linearization. This simple technique considers small perturbations about an operating condition so that nonlinear gain curves can be represented by a constant equal to the slope of the curve at that point. Also, functions such as multiplication, division, square rooting, etc. can be replaced by the summation of the partial derivatives of the output for each input considered separately. Analysis is then reduced to the familiar linear methods already covered.

Discontinuous nonlinearities are more difficult to analyze since they result in system behavior that varies not just with frequency but also with signal amplitude. The hysteresis nonlinearity was singled out as the most troublesome due to the fact that it can generate phase shifts of up to 90° and is often the source of limit cycling types of instability.

The describing function approach to determining whether or not a specific nonlinearity can result in instability is somewhat limited since it can only provide a 'yes' or 'no' answer. A far more useful method for evaluating the effects of nonlinearities is to use simulation and modeling techniques. This approach takes care of all types of nonlinearity and can provide the analyst with unlimited dynamic performance information. It is important, however, to have some understanding of what to expect from simulation exercises and to question unexpected results until an explanation is found.

Chapter 6 addressed the use of the digital computer as a controlling element in closed loop systems since this is becoming more the norm with the continuing improvements in the cost and speed of electronics. Here we learned how to take into account the effects of signal digitization and re-conversion and the time delays associated with these processes.

The section which follows will attempt to capture the most important rules, and procedures that have been covered in the book as a whole. References to the specific section(s) are also included to assist the reader.

7.2 Graphical Tools

The use of graphical techniques has been emphasized throughout as a powerful supporter of the analytical process because they are easy to use and provide good visibility as to what is happening dynamically with the control system under scrutiny. The Bode diagram (frequency response plot) is by far the most popular graphical tool used by the control engineering community and is ideal for showing the degree of stability, in terms of gain and phase margins, that can be expected from a closed loop control system.

The first thing to remember is that all systems can be represented by transfer functions that are combinations of first- and second-order elements whose response characteristics, in terms of gain and phase, are well documented as tables or graphs expressed as ratios of frequency with either the time constant or the undamped natural frequency. Thus it is an easy task to translate each problem-specific element into gain and phase plots to obtain the composite open loop gain of the system. Using the frequency response graph this is made simple by the fact that the gain asymptotes for the dynamic elements are simple straight lines. For example:

- first-order elements have a flat gain response up to the break frequency $1/T$ and thereafter the gain attenuates at a constant rate of $6.0\,dB$ per octave ($20\,dB$ per decade);
- second-order elements have a flat response up to the undamped natural frequency ω_n and thereafter the gain attenuates at $12.0\,dB$ per octave ($40\,dB$ per decade). The resonance effect around the natural frequency is a function of damping ratio and can be easily estimated from standard curves.

The phase angle for each element, however, is not as easy to define since it does not change linearly with log frequency and the curves for each element must be sketched with the help of a few 'magic numbers' that we can easily commit to memory. Again the total phase is simply the sum of the individual elements' contributions.

The 'short cut' approach to stability assessment described in Chapter 2 promotes the use of simple rules using only the gain plots to determine the acceptability (or otherwise) of the stability margins of the system. While this approach is only approximate and perhaps a little conservative, it is an easily used method that can provide a quick assessment of the situation. This method states that using only the open loop gain versus frequency plot, good stability will result if the gain curve crosses the zero dB line with a slope of 6.0 dB per octave for about half a decade either side of the cross-over frequency. This approach eliminates the need to generate the phase angle plots which is perhaps the most tedious chore in the analysis process. Figure 7.2 shows examples of this method showing a system that will exhibit good stability and one which would probably have unacceptably small stability margins. Another useful guideline in closed loop control system design is that it is good practice to ensure that there is good separation between the bandwidth of the process being controlled and the bandwidth of

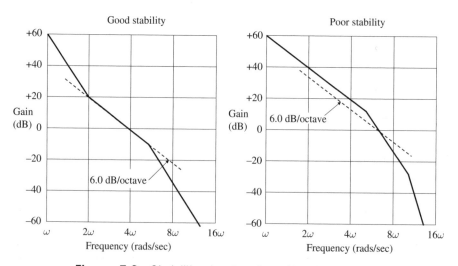

Figure 7.2 Stability short cut method example

the controller. Also it is desirable that feedback sensors should be as fast as possible so that their contribution to the system dynamics is minimized.

7.3 Compensation Techniques

Chapter 3 covered the basics of control system compensation providing detailed descriptions as to how compensation transfer functions can be built to suit the specific needs of the system under review. This was reinforced through the use of several design examples. There are a number of key lessons learned which should be passed on to prospective control engineers that may prove useful in future exercises. The following is a short list of system features related to the compensation development process that are worth noting.

7.3.1 Integral Wind-up

We have used the integration process as a means to eliminate errors in steady state and the use of compensation techniques to minimize the effect of the unwelcome phase lag that accompanies the integral action. We also need to be aware of how the integrator responds following large changes in the control loop command. During large transients, as long as the error input to the integrator remains finite and of the same sign, the output from the integrator will continue to increase to a point where it is caught with a large output as the process reaches the commanded value and the error changes sign. In the time it takes for the integrator output to move back towards its null position the process output can exhibit a large overshoot beyond the commanded value.

This phenomenon is known as 'integral wind-up' and can be controlled by limiting the authority of the integrator. This can be accomplished as indicated in Figure 7.3 where a high gain feedback is introduced around the integrator which comes into play as the authority limit is reached.

In today's modern digital controls with the flexibility provided by software this is even easier. For example, the integrator gain can be made nonlinear with additional logic to control when the integrator is operative or not as a function of the error.

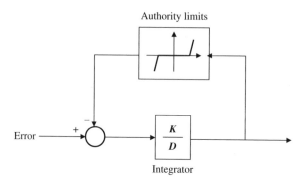

Authority limits

Error

Integrator

Figure 7.3 Integrator with authority limits

7.3.2 Avoid Using Pure Derivative Action

Derivative action is very attractive to the control system designer because, in theory, it provides a 90 degree phase lead right off the bat! We need to recognize, however, that the differentiation process is inherently noisy and should only be employed with an attendant high frequency filter term to avoid the magnification of noise. The pure derivative transfer function TD is not recommended. The pseudo derivative:

$$\frac{T_1 D}{(1 + T_2 D)}$$

is much better. This approach was covered in Chapter 6 on digital electronic control where the generation of a pseudo derivative algorithm was described. In the preferred transfer function, the lag term effectively cancels the derivative term for frequencies above about 10 times the break frequency, i.e. $10/T_2$ radians per second.

7.3.3 Mechanical Stiffness Estimates are Always High

This point was made more than once in Chapter 3; however, it is an important message that cannot be overstated. All too often the control system designer has to contend with lower resonant frequencies resulting from low stiffness estimates. It therefore behooves the designer to consider at the outset what might be done to compensate for such an event. Better yet, use test results where possible to define the best control action.

Finally, on the general subject of compensation, the control system designer can gain significant insight into the effects of adding various control elements into the loop by using the complex frequency domain (the '*s*' plane) to observe the location of the open loop roots and to sketch in the root locus curves with and without the compensation. Further discussion of this subject is presented in the following section.

7.4 Laplace Transforms and Root Locus Techniques

Laplace transforms gave rise to the application of root locus theory which became very popular in the aerospace industry in the 1960s and 1970s. The aircraft as a dynamic process behaves in a linear fashion for small deviations about a given flight condition. This type of process lends itself well to the use of root locus and this became the tool of choice for the design and analysis work associated with aircraft stability and autopilot systems. The example in Figure 7.4 shows how an aircraft that is basically unstable in pitch can be stabilized via a simple pitch rate control loop.

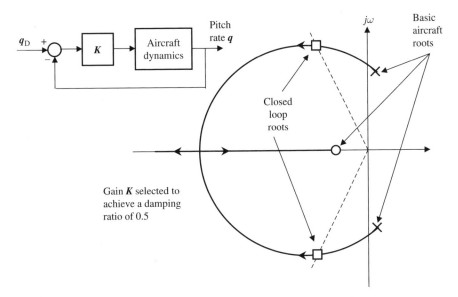

Figure 7.4   Root locus example for a pitch axis autostabilizer

The aircraft dynamics are represented by the following transfer function which comprises a first-order lead term in the numerator and a second-order term in the denominator:

$$\frac{20\,(s+2)}{\left[s^2 + 2\,(-0.1)\,(4)\,s + (4)^2\right]}.$$

The latter term shows the undamped natural frequency to be 4.0 radians per second with a damping ratio of -0.1 hence the location of the aircraft roots on the positive side of the $j\omega$ axis.

In this example, the root locus plot can be quickly sketched and the gain required to provide the desired damping ratio obtained. In the past, slide-rule type tools (called 'spirules') were available to help the analyst to add and subtract angles to locate the -180 degree locus on a scaled graph. This same tool could be used to calculate the gain at any point on the locus. Today this can be accomplished using readily available software tools that will run on standard PCs; however, the purpose of introducing the root locus method here is make the reader aware of its capabilities and to encourage its use from a qualitative perspective. This again provides valuable insight into a control system's behavior from the location of the open loop roots to the location of the closed loop roots over a wide range of loop gains. It also gives the analyst useful information as to what the various compensation elements will contribute to the potential closed loop performance.

The advice, therefore, is to use Laplace transforms and root locus as another tool that can provide additional visibility into control system behavior. Having obtained the open loop transfer function, it is easy to sketch out the root locus plot to see what is happening in the complex frequency domain and where the loci track as gain is increased.

7.5 Nonlinearities

We must remember that in the real world purely linear systems do not exist and that the control systems engineer must use linearization techniques that provide ease of analysis while maintaining a reasonable representation of the fundamental dynamic behavior of the system being analyzed. Fortunately most feedback control systems can be adequately studied using the basic linear analysis and synthesis techniques described in this book; however, it is always a good idea to keep asking the question 'are my assumptions sufficiently valid?'. Experience suggests that the problems with performance in the field are primarily due to the fact that the real world use does not adequately match the intended application. This suggests that we need to be much more attentive to the way products are used than blindly referring to the specified performance requirements.

As an example let us consider the hydraulic servo actuator with overlap in the servo valve spool. This feature manifests itself as a deadband around the null point so that the flow gain of the spool valve is reduced for small inputs to the spool valve. From the perspective of actuator stability, this is not particularly significant since it means that for small signals the loop gain is progressively reduced as input amplitude is reduced. Figure 7.5 shows typical test results that can occur from this type of situation. As shown in the figure, the response degrades as the amplitude of the input command is reduced. This degradation manifests itself as a small reduction in gain and, more importantly, as an increase in phase lag and while the actuator itself is quite stable its degraded performance at low amplitudes can seriously impact the

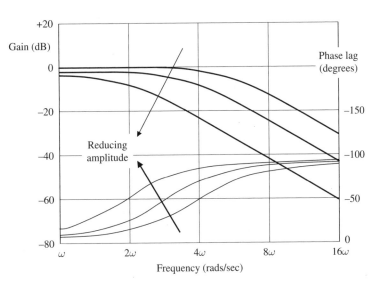

Figure 7.5 Degraded actuator response due to valve overlap

stability margins of an outer control loop of which the actuator is just one element.

It is therefore important for the design engineer to pay attention not just to performance aspects of one component when it plays a role as an element in a larger system. Component specifications are notorious for being incomplete and, in the complex integrated systems environment that we find ourselves today, every engineer should be prepared to share an understanding of the big picture early in the design and development phase in order to minimize the need to fix performance problems after the system enters service.

7.6 Digital Electronic Control

The electronic control chapter (Chapter 6) provided the reader with a basic understanding of how a digital computer can be utilized as a controller in a closed loop system. In today's environment even mechanical engineers need to have an understanding as to how signals are digitized (and vice versa) and the time delays involved in the data conversion and calculation processes in order to appreciate the impact on closed loop stability that these characteristics generate.

Once again this stresses the need for today's engineers to be not only specialists in their specific field but also generalists who have a fundamental appreciation of the pitfalls that can trap the unwary or uninformed engineer who does not have a feel for the 'big picture'. In order to simplify the task of the typical digital controller it was recommended that the operational software program be considered as two subsections that together perform the total controller function namely:

* the input/output (I/O) handler;
* the control logic.

The I/O handler focuses on the conversion of sensor data into the digital domain and performs built-in-test (BIT) checks on all converted data to ensure that the input signal conditioning circuitry and output drivers are operating correctly, while the control logic is concerned only with the overall system functionality.

A consideration in the design of the software architecture is to ensure that the execution is performed in a deterministic fashion in order to provide an operational environment that is amenable to the establishment of a predictable verification testing regimen as part of the certification process. Architectures that involve layers of interrupts with different levels of priority should be avoided since the execution of the software program becomes indeterministic thus making verification testing and regression testing following changes and problem fixes to often be inconclusive. Even more seriously, the nature of indeterministic software is its ability to hide embedded problems during the development and certification phase of the program only to have them suddenly appear well after the system enters service when a much longer software operational exposure time has occurred.

Another architectural design issue to be aware of concerns the use of multiple control channels executing the same control logic program for the purpose of providing control redundancy. In the event of a failure in one control channel, the remaining channel (or channels) can maintain continued safe control of the process. Problems can arise, however, when the control channels are not synchronized in time. In this case even minute differences in clock frequency can result in two (or more) channels becoming a full clock cycle out of step and, as a result, the redundancy management logic whose job it is to monitor differences between channels can erroneously de-select a healthy channel. While some of the above commentary is only secondarily related to the feedback control issue it is considered to be of sufficient importance to be brought to the attention of the prospective control systems engineer.

7.7 The Way Forward

The content of this book addresses only a small corner of the subject of feedback control systems engineering within an area that is referred to as classical control theory. Hopefully the reader has found the material to be both relatively easy to absorb and interesting. As an introductory book there are a number of the less commonly used topics associated with the classical theory that were not covered here including the following.

(a) *Random noise.* This approach to both analysis and testing utilizes the concept whereby random noise comprising a specific power spectrum and frequency content is used as the input to closed loop control systems. It is interesting to realize that if a system is excited with random noise containing a mixture of all frequencies, the system will act like a tuning filter by magnifying those frequencies that it 'likes' while rejecting all of the other frequencies. Thus the transfer function of a system tested using random noise can be synthesized via this technique.

(b) *Phase plane analysis.* The phase plane is a graphical method of analyzing the dynamic behavior of graphs of second-order systems when nonlinearities exist that can be expressed as functions of output velocity and position but are not time dependent. Systems with on–off (bang–bang) controllers are an example where the phase plane method can be used effectively. The approach is to redefine the equations of motion as functions of velocity and position and to develop graphs of these two variables plotted against each other. These response curves define the system response in the 'phase plane'.

(c) *Sample data systems.* This is the rigorous analysis method for evaluating the stability of sampled data control systems and involves another mathematical transform technique (the 'Z' transform) that can be used in block diagram form as well as via the 'Z' plane chart to provide insight into their functional behavior.

Now that we have successfully penetrated the mathematical mystery of feedback control theory, it should be relatively easy to broaden ones knowledge through further reading. There are many books on classical control theory available today with different areas of emphasis and the reader is encouraged to seek out what is most appropriate in terms of the industrial area of interest, style and material content.

Beyond classical control theory, which represents the limit in scope of this book, there is modern control theory that applies matrix mathematics in the exploration of the dynamic behavior of multi-input, multi-output systems. This area of study is the cutting edge of control theory and continues to be the subject of most advanced degrees in the field of control engineering.

Index

Stability and Control of Aircraft Systems: Introduction to Classical Feedback Control   R. Langton
© 2006 John Wiley & Sons, Ltd